Jet Fighter School II
More Training for Computer Fighter Pilots

Richard G. Sheffield

COMPUTE! Books
Radnor, Pennsylvania
Greensboro, North Carolina

Other books by Richard G. Sheffield:

Jet Fighter School: Air Combat Simulator Tactics and Maneuvers
Sub Commander: Tactics and Strategies for World War II Submarine Simulations
Gunship Academy: Tactics and Maneuvers for Attack Helicopter Simulations

Cover photograph courtesy of SubLOGIC Corporation.
Maps in section on F-15 Strike Eagle courtesy of MicroProse Software, Inc.
Interior photographs courtesy of the National Archives

Editor: Robert Bixby
Illustration by Adrian J. Ornik

Printed in the United States of America

10 9 8 7 6 5 4 3 2 1

Library of Congress Cataloging-in-Publication Data
Sheffield, Richard G.
Jet fighter school II.
Includes index.
1. Computer war games. 2. Fighter plane combat—Computer simulation. 3. Stunt flying—Computer simulation. 4. Flight simulators. I. Title. II. Title: Jet fighter school 2. III. Title: Jet fighter school two.
U310.S46 1988 358.4'148 88-29917
ISBN 0-87455-157-9

COMPUTE! Books, Post Office Box 5406, Greensboro, North Carolina 27403, (919) 275-9809, is a Capital Cities/ABC, Inc. company and is not associated with any manufacturer of personal computers.

Contents

Foreword

Nothing can be more thrilling than aerobatic flying performed by a seasoned professional, unless it's flight in combat. You'll find both in generous supply in *Jet Fighter School II: More Training for Computer Fighter Pilots*.

Learn Aresti, the language of aerobatic flying, and develop creative aerobatic routines of your own, composed of the many maneuvers detailed in this book. Or take advantage of the new and exciting missions developed by the author for *F-15 Strike Eagle*. It's all here.

Learn barrel rolls, figure eights, Immelmanns, and chandelles: maneuvers difficult and beautiful in their own right (they've been drawing crowds to air shows for decades), but also useful in combat. Learn to fly simple maneuvers, combine them into complex routines, and finally, learn to combine your flying with another plane to produce such eye-catching formations as the head-on and tandem passes, and the famous bomb burst.

Whether you're strafing surface-to-air missile sites in Iran or bombing Tripoli in retaliation for the latest horrendous terrorist attack, you'll earn your wings every time you take to the sky. Learn the most strategic payload to carry, and how to make the best use of your payload. Learn defensive measures that can enable you to outfly a heat-seeking missile in close quarters.

Experienced simulator pilot and noted author Richard Sheffield fills you in on the strategies that will keep you flying even in the face of superior forces. He's done the homework to provide you with the adventures of a lifetime.

Introduction

This book is divided into three parts that correspond to the approach I take to playing flight simulation games.

First, I look at the quality of the simulation itself: How accurately does it simulate the flight characteristics of the aircraft involved? Does the graphic environment work? Are you presented with enough visual cues to let you fly the aircraft in a normal manner without over dependence on the instruments? A high-quality simulation should let you fly a variety of complex maneuvers independent of the game aspect of the simulation.

Part 1 of this book examines a number of simple and complex aerobatic maneuvers you can learn and link together. This type of flying is independent of the normal operation of the simulation as a game, but will help you master the aircraft and push the capabilities of the simulation to its limits.

Part 2 deals with the second aspect of most of these games: combat. This section deals exclusively with one of the most popular air combat simulations, *F-15 Strike Eagle.*

A series of training missions is described. These missions are designed to simulate the Air Force training exercise Red Flag. Once you have completed your training, you can move on to four new action-packed mission scenarios that use the existing maps to simulate real and possible future combat situations.

Finally, Part 3 deals with a process we all go through when playing these games: the process of strategic and tactical planning. Most of us take many missions to figure out what works and what doesn't. Part 3 includes strategic tips and playing hints for many of the most popular games currently on the market. These tips are designed to get you through the frustrating initial period of game play, so you can spend less time trying to figure what is going on and more time playing and enjoying the game.

I hope I've included something of interest to every computer pilot.

Part 1

Introduction to Aerobatics

CHAPTER 1
The History of Aerobatics

A common notion is that aerobatics and stunt flying were developed during World War I when planes were first used in combat. This is far from the case. The history of aerobatic flight significantly parallels the history of flight itself.

The Origins of Aerobatic Flight

The Wright brothers themselves performed the first aerobatic maneuver (a *360-degree banked turn*) in September of 1904. The key word here is *banked* because this was prior to the invention of the aileron. The Wright flyer used a *wing warping* system to flex the entire wing in order to bank the aircraft.

The whole idea of banking the aircraft greatly speeded up the development of flying machines in the United States. The Europeans considered the idea unsafe and unnecessary. They tended to view the airplane as a flying automobile which could be steered around without tilting the wings: an idea that limited their progress for a number of years.

While the Wrights were perfecting their flight control system, even better systems were being developed elsewhere. Dr. William Christmas in Washington, D.C., and a group headed by Glenn Curtiss and Alexander Graham Bell, almost simultaneously developed the *aileron*.

The Curtiss machine featured an aileron control surface mounted between the biplane wings, close to the tips. Dr. Christmas, however, had a better idea and integrated the aileron as a hinged section in the wing itself—a concept still used today. While Curtiss and the Wrights argued over infringement of the Wright's wing warping patent, Dr. Christmas quietly received a patent for his design. In fact, congress later paid him $100,000 to compensate him for the use of ailerons in all aircraft built during World War I. I mention the aileron because it is the main development that led to more complex aerobatic maneuvers.

By 1909 the Europeans were starting to catch up. In July of that year, a Frenchman, Louis Blériot, made the first flight across the English Channel. This sparked a widespread interest in flying throughout Europe and America. Flying shows and air meets became wildly popular, and highly profitable for the pilots. Large paying audiences gathered wherever the flyers performed, but they soon tired of simple flying exhibitions and demanded thrills and danger. To keep the crowds coming, the pilots competed vigorously to develop flying tricks and stunts: today's aerobatic maneuvers.

Figure 1-1. A Blériot Monoplane of the Type Used in the First Flight Across the English Channel

Daredevils

An American, Walter Brookins, captured the attention of audiences with his *spiral dives* and *90-degree banked turns,* still considered wild and dangerous in 1910. Lincoln Beachey countered with his *Death Dip,* in which he would fly to 5000 feet and dive straight at the ground with the engine off. At the last second he would pull up, often causing quite a stir in the audience as he passed them with both hands off the controls! Beachey loved to scare an audience and would often fly through a stand of trees or under telegraph wires.

As wild as the pilots were during this period, the aircraft often couldn't keep up. Flyers were quickly passing the structural capabilities of the airframe, often with fatal results. In a six-week period in 1910, three of the most famous pilots of the time, Ralph Johnstone, Arch Hoxsey, and John Moisant, were killed when their airframes collapsed while pulling out of diving maneuvers. The press was extremely hard on Beachey and blamed him for many of the deaths because pilots tried to match his skill. They went as far to say that whenever there was a fatal accident, the pilot had *pulled a Beachey*. Distraught by this and the death of a close friend and team member, Beachey went into temporary retirement in 1912.

Flying at slow speeds resulted in constant danger of a *stall* and a *spin*. The result was often fatal because no one had yet discovered how to recover from a spin. Such was the situation presented to Wilfred Parke in August of 1912. He fell into a left-hand spin during a military test. After pulling hard on the stick and pushing the rudder to the left with no result, he eased off the rudder and pushed it to the right, into the spin. The plane immediately righted itself with about 50 feet to spare.

Parke was a detail-oriented test pilot and immediately analyzed and wrote down his experience. Now that someone had entered a spin and lived to tell how to correct it, the mystery of the spin was finally solved. It would take years for the word to spread however, because Parke died shortly thereafter. Eventually the spin would become part of the aerobatic bag of tricks.

Upside Down and Backwards

By 1913, the Europeans were progressing nicely with their French Blériot and Morane monoplanes and rotary engines. No longer satisfied with normal flight, they began to experiment with unusual flight characteristics. An Englishman, Will Moorhouse, was the first to fly a plane backwards. He pulled up into a steep climb, pulled the nose up as far as it would go, and killed the engine. The Blériot-type monoplane stopped for a moment and then slid backwards for a short distance before yawing to one side and diving nose-first towards the ground. This later became a standard maneuver know as the *stall-turn*.

Inverted flight was also being considered. Several pilots had inadvertently found themselves upside down as a result of wind gusts, but no one had yet attempted it intentionally. Adolphe Pégoud decided he would be the one to try.

Pégoud was one of the first true aerobatic pilots. An accomplished test pilot with the Blériot team, he was also the first flyer to jump from a plane with a parachute. Both Pégoud and the aircraft landed, unharmed. Having decided to try upside-down flying, Pégoud first practiced in a hanger in a plane hung upside down from the ceiling. He correctly assumed the controls would have to be operated in a reverse manner and he wanted to get the feel of it before actually flying.

It must have worked. Shortly thereafter, Pégoud made the first public demonstration of inverted flight in September of 1913. During this flight, he was the first to perform a *half roll* to an inverted position—another aerobatic maneuver was born. The flight did point out one potential problem: While Pégoud was prepared to fly inverted, the aircraft was not. It proceeded to drench the flyer with fuel!

Until that point the glories of flight had been limited to the Americans and Europeans. In September of 1913 another player entered the picture. A Russian pilot named Petr Nikolaevich Nesterov performed the first *complete loop*. The first reaction of his superiors was to arrest him for taking undue risks with equipment. Shortly they reconsidered and promoted him.

The name of Petr Nikolaevich Nesterov would surface again in the opening months of World War I. At this time, he was in charge of an air squadron fighting the Germans. When a German plane was spotted over the field, Nesterov ran to his plane and took off unarmed. He approached the German Albatross airplane and tried to destroy it with his landing gear. He misjudged the approach and rammed the German with the front of his monoplane. During the uncontrolled spin to the ground, Nesterov was thrown from the plane. He and the German crew were all killed. Nesterov went down in history as one of the first fighter pilot heroes. He was buried with full military honors.

Suddenly, a looping mania took over. Everyone started doing loops. The loop is even credited with bringing Lincoln Beachey out of retirement. He had Glenn Curtiss design a new plane for him and went on tour performing loops, *barrel rolls, spiral dives,* and probably many more maneuvers that weren't recorded.

Beachey quickly became the king of the loopers. He was looping all over the country and charging by the loop. He made $500 for the first loop and $200 for each loop thereafter. In 1914 he was earning $4000 to 5000 per week. He frequently flew loops in sequences of ten or fifteen at a time, starting at less than 1000 feet. He was probably also the first person to perform an *outside* (inverted) loop.

Figure 1-2. Lincoln Beachey, the King of the Loopers, Upside Down

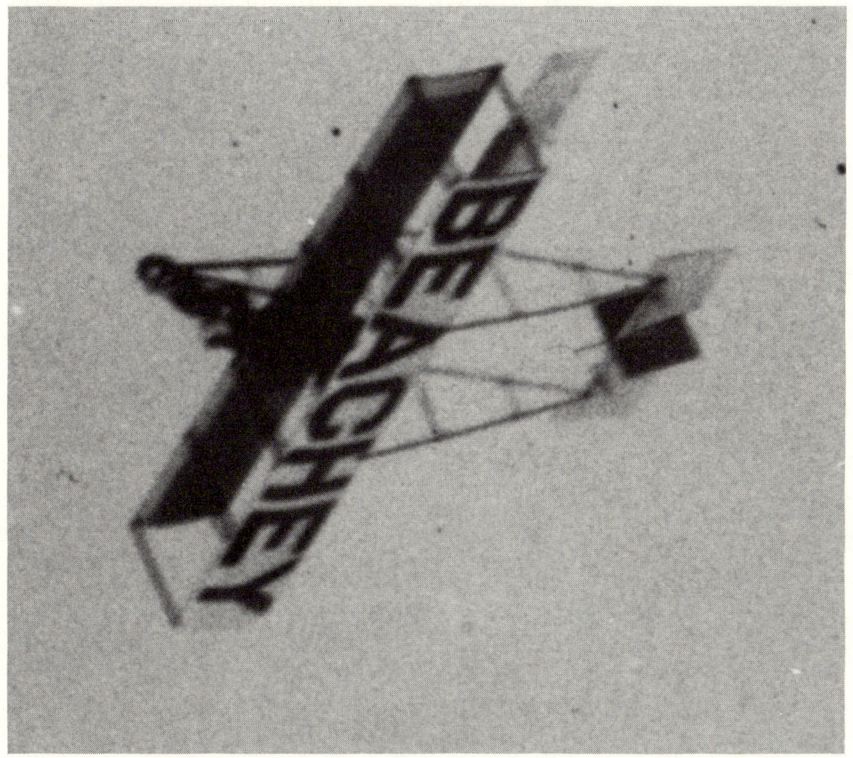

Beachey also worked in aircraft design, developing a sleek monoplane. On March 15, 1915 he unveiled his new design at an exhibition in San Francisco, his home town.

After performing a series of loops, he pulled into an *S-dive* over the bay. Having flown biplanes for a number of years, he apparently was taken by surprise by how fast the new monoplane dived. He quickly found himself running out of altitude and pulled back hard on the stick. The plane was never designed to perform this maneuver at such high speeds. The crowd stood dumb-struck as both wings snapped off and Beachey plunged to his death into the bay. Orville Wright referred to him as "the greatest aviator of them all."

World War I

By 1914, the word *aerobatics* was a part of the language both in the U.S. and abroad. Air shows and flying exhibitions were frequent and heavily attended. The sport was healthy and growing prior to the first shot of World War I.

The Great War had enormous influence on aerobatics, as aerobatic tricks and stunts quickly became lifesaving maneuvers in battle.

Two innovative Germans, Oswald Boelcke and Max Immelmann, made their mark in the sport during this period. Boelcke was the master tactician and leader; he developed many of the tactics used so successfully by the German aviators and has been called the "father of air combat." Though the Germans were flying the somewhat inferior Fokker Eindecker monoplane, their superior tactics gave them an early advantage over the battlefield.

Figure 1-3. Oswald Boelcke, Father of Air Combat

Max Immelmann was known for his successful hit-and-run attacks. He was a master of the surprise attack, often coming out of the sun or from underneath his opponent and then using aerobatic maneuvers to get away. His name lives on in modern aerobatics in the form of the *Immelmann Turn,*

an ascending half loop with a half roll. There is a controversy, though, as to whether this maneuver could have been flown in the wing-warping Fokker monoplane. Immelmann is still given credit for using this maneuver to attack from beneath an opponent on the way up, then reverse course and get away. It is more probable that the maneuver used was similar to the modern *wing-over* or *chandelle*.

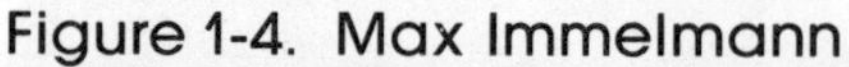
Figure 1-4. Max Immelmann

The war also took its toll on a number of great flyers. The great Adolphe Pégoud was shot down and killed in August of 1915, only two years after he pioneered inverted flight. Oswald Boelcke died in a midair collision in 1916. By the end of the war, a number of new tricks had been added to the aerobatic inventory. *Half loops, barrel rolls,* and the *split-S* all became commonplace due in part to the development of the powerful German Fokker Albatross and the English Sopwith. The war also gave aerobatics a valid reason for existence; those pilots who lived long enough to develop their aerobatic skills over the battlefield were the most successful. Despite the obvious importance of this type of flying, formal training courses for the military were not set up until 1917.

Figure 1-5. The Feared German Fokker Albatross

After the Armistice

With the war over, hundreds of flyers had nowhere to fly. The reappearance of the air show was soon to follow. As before the war, a competitive spirit set in among the performers, each trying to outperform the rest. Some of the shows featured the famous barnstormer acts while others pursued the more traditional aerobatic displays. Jimmy Doolittle, who later bombed Tokyo, pioneered inverted flight and revived interest in the *outside,* or *negative G loop.* The intense rivalry among the flyers continued to produce new maneuvers. *Vertical rolls, flat spins,* and *vertical figure eights* were added to the list as well as the *roll on the top of a loop,* or *avalanche.*

This period also featured the first aerobatic competitions. The world's first large scale international aerobatics competition was held in Zurich in August of 1927.

A young German pilot named Gerhard Fieseler had developed such a reputation that he was invited to the Zurich meet. Fieseler wanted to do well and was in search of new maneuvers when the idea of inverted flight became popular again. Several months of intense work produced the first aircraft fuel system capable of sustained inverted flight. Fieseler quickly mastered a series of inverted flight figures which could be flown in any weather.

From the start of the competition it was a three-way race between Fieseler and two French pilots. After a series of compulsory figures and short free-style programs, Fieseler was still in the hunt. He progressed to the finals where he executed his unique and flawless routine. Fieseler was awarded only second place, however, perhaps for political reasons. Whatever the reason, inverted flight had arrived and would become a permanent part of aerobatics.

Competition aerobatics became a mature sport complete with rules and regulations. International meets became quite common. In the mid 1930s a new force in world competition would emerge from Czechoslovakia. The Czechs had been developing their flying skills since the days of Pégoud and were now ready to show them to the world. They won two of the top three spots in the 1936 Olympic competition and almost all of the titles in Zurich the next year. They were undeniably on a roll. But that roll was stopped cold by World War II.

World War II and Beyond

World War II saw a dramatic increase in the power and speed of fighter aircraft. Due to these high speeds, many of the aerobatic maneuvers of the World War I became obsolete and dangerous. Few aerobatic maneuvers came out of the conflict but a number of important aircraft modifications did result. Seats were modified to help the pilot withstand higher G forces. Straps were placed on the pedals to help keep the pilot's feet from sliding off during maneuvering. The German Stuka dive bomber canopy included marks to identify the plane's dive angle and a window was placed in the floor of the cockpit. All of these changes made their way into postwar aerobatic aircraft.

In the 1950s, just when everyone thought that all possible aerobatic maneuvers had been discovered, the Czechs reestablished their leadership in the sport by introducing the *lomcovak*. This maneuver must be seen to be understood.

Even then it's difficult to comprehend, as many of the first observers found when they tried to repeat what they had witnessed. The maneuver is best described as a series of gyroscopic twists and somersaults during which the plane rotates about all three axes.

Aircraft design improvements continued through the following decades, allowing a few more innovations (mainly in the area of vertical maneuvers) as increased power and lighter aircraft made vertical acceleration possible. The recent emphasis, however, is on the accurate performance of the maneuvers; loops in perfect circles and accurate figure eights have taken the place of dangerous thrill-seeking maneuvers, much to the benefit of the sport and the pilots involved.

As you can see, the history of aerobatic flight is as old and honored as the history of flight itself. Through the years it has been the realm of a few high-spirited and courageous individuals and out of the reach of most people. Now anyone with a computer can share in the experience of a well performed maneuver and in a small way identify with those daring men who pioneered the sport in aircraft made of little more than cloth, wood, and wire.

CHAPTER 2
Basic Maneuvers

If you've ever seen expert pilots perform aerobatic routines at an air show, the thought of flying these maneuvers yourself may seem a bit overwhelming. It need not be. With computer simulations you can fly any routine as well as the experts—maybe even better. The next time you see the Blue Angels or the Thunderbirds, take note of their routine. You'll quickly recognize that the most complex shows are made up of a number of simple maneuvers skillfully laced together.

The purpose of this chapter is to teach the basic maneuvers. These maneuvers are the building blocks for more complicated tricks. When you perform a number of these maneuvers back-to-back you have the makings of an air show routine.

It's important that you take the time to master these basic flight patterns. Make sure you can perform each maneuver smoothly with a controlled entry and exit, especially the exit. Once you master these maneuvers you can begin to put them together in a sequence. But you won't be able to make the transitions from one trick to the next unless your exit from each maneuver is controlled. If you don't have control coming out of a trick, you won't be able enter the next maneuver with your wings level and the nose pointed in the right direction.

When learning these maneuvers give yourself plenty of altitude at first. Having to reboot after a crash is frustrating while it slows the learning process. Also, try to practice with landmarks in mind so you can perform your maneuvers in the right direction and over the correct spot on the ground. Remember, you'll be performing for an imaginary audience on the ground. Stay close by so they can see you.

As mentioned, each simulation is different, so take note during your practice of how your plane reacts in each maneuver.

- How much altitude do you lose?
- How much do you have to anticipate to get the wings to stop level?
- How much speed do you loose, or gain?

⊙ Do you come close to stall speed?
⊙ How much does the plane change course during a roll?

You need to have an intuitive feel for the aircraft. You don't have all the sensory inputs that a pilot has, such as G forces and aircraft vibrations, so you must pay close attention to the instruments during the early phase of aerobatic training. Several of the simulations offer a feature that allows you to turn off some of the ground scenery and increase the animation speed. If the game you're using offers this feature, by all means use it when performing aerobatics. Your plane will react much faster to your control inputs. This will make your maneuvers smoother and more precise.

Try to fight the urge to move quickly to the more flashy and complicated maneuvers. Taking the time to master basic skills will pay dividends down the road.

The Basic Aerobatic Maneuvers

Each of the maneuvers described below has an accompanying illustration; the numbers in parentheses refer to the numbers in the drawing; each number indicates an action to be taken during the maneuver.

When you see the expression *apply full forward stick* or *give some right stick,* this means to push the joystick in the direction indicated with the intensity indicated. This is not common aeronautic parlance, but it's useful to have a shorthand expression because flight simulators are joystick-intensive; using more elegant language to describe joystick movement would add to the bulk of descriptions without making them clearer.

The Loop

The loop is one of the most important aerobatic building blocks. Although it looks easy and can be performed by simply pulling back on the stick and holding it, this will not produce a round loop. Your goal is to perform a nearly perfect circle and exit this maneuver at the same altitude and on the same heading you entered.

⊙ All stick maneuvers for the loop are gentle and gradual.
⊙ Fly straight and level at approximately 500 knots. Start pulling back on the stick a little at a time. (1)

- ⊙ As you get close to vertical, apply some forward stick. Your goal is to wind up in level, inverted flight at the top of the loop. As you reach the inverted position, repeat the process of pulling back on the stick. (2)
- ⊙ When you get to a position where your nose is pointing straight down, start adding some forward stick gradually so you have a nice, gentle pull out with your wings level. (3)

Performing the loop in this fashion will allow you to fly a round loop rather than an egg-shaped loop with a ballistic climb and dive.

Figure 2-1. The Loop

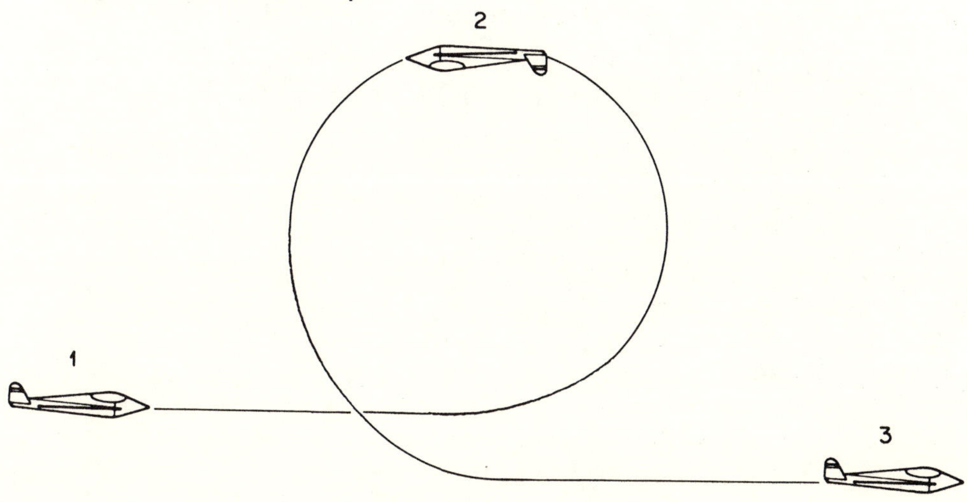

The Aileron Roll

The aileron roll is used as a part of many other maneuvers and it's quick and easy to learn. You should try to maintain the same heading and the same altitude when you exit as when you started the maneuver. With some simulations, the aircraft rolls so slowly that your heading changes considerably. It should, however, balance out at the end of the roll.

- ⊙ Fly straight and level at a speed greater than 300 knots. (1)
- ⊙ Push the stick all the way to the right or left; be careful not to move the stick forward or back at the same time you push it to the side. (2)
- ⊙ The horizon should start to spin around. (3)
- ⊙ As you approach level flight again, ease up on the stick. (4)
- ⊙ Try to stop the rotation with your wings perfectly level. (5)

You should practice all roll maneuvers both to the right and the left to avoid becoming right- or left-handed.

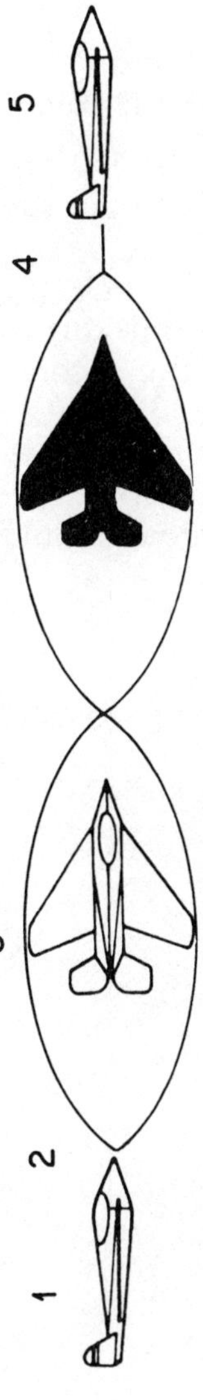

Figure 2-2. The Aileron Roll

The Barrel Roll

The barrel roll is often used in aerobatic demonstrations as well as in air combat. It's halfway between a roll and a loop.

- Fly straight and level at 300 knots or more. (1)
- Pull back on the stick to get the aircraft into a 30-degree climb. (2)
- Apply full right or left stick. (3)
- The aircraft should climb during the first half of the roll and descend during the second, or inverted portion of the roll.
- As your wings approach a level position, ease up on the stick and apply a little forward stick to end the maneuver in level flight. Your goal should be to exit at the same altitude you entered. (4)

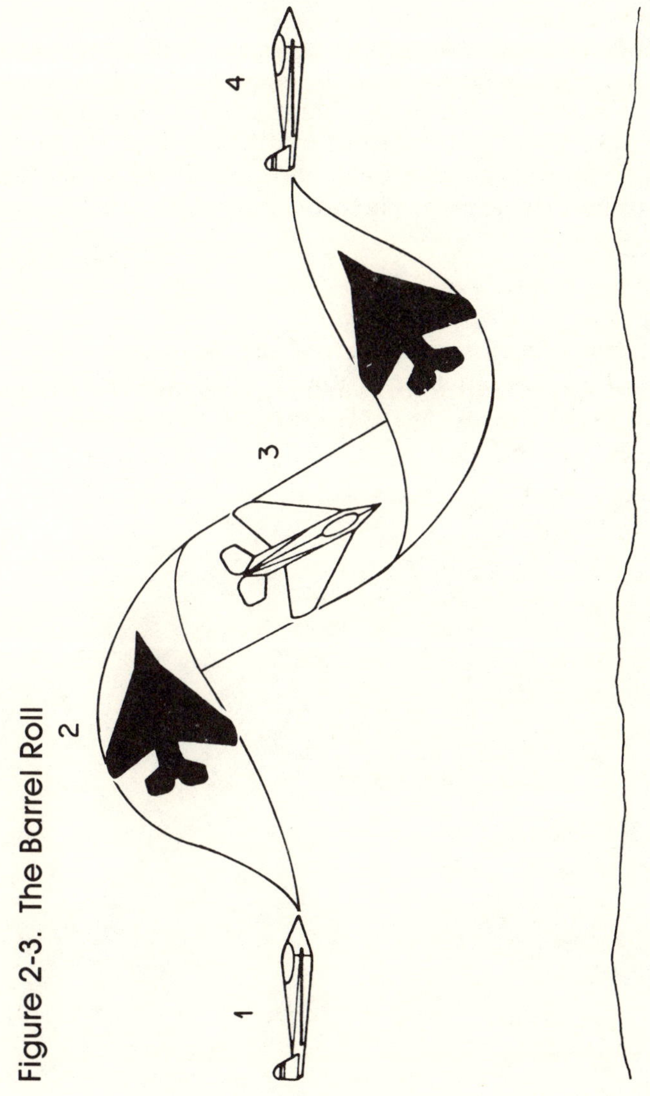

Figure 2-3. The Barrel Roll

The Split-S

The split-S, or descending half loop, is often used to transition from a high altitude maneuver to a low altitude maneuver. It is also used as a defensive maneuver and a bombing method in combat. The minimum altitude for this maneuver depends on the load you're carrying and the amount of thrust applied during the descending portion of the maneuver.

⊙ Fly straight and level. (1)
⊙ Apply full right or left stick. (2)
⊙ Stop the rotation in an inverted position. (3)
⊙ Pull back on the stick and hold it until the horizon appears. Pull out in level flight. (4)

Once you've mastered the split-S with this method, try pulling back on the stick during the half roll portion of the maneuver to roll and dive at the same time.

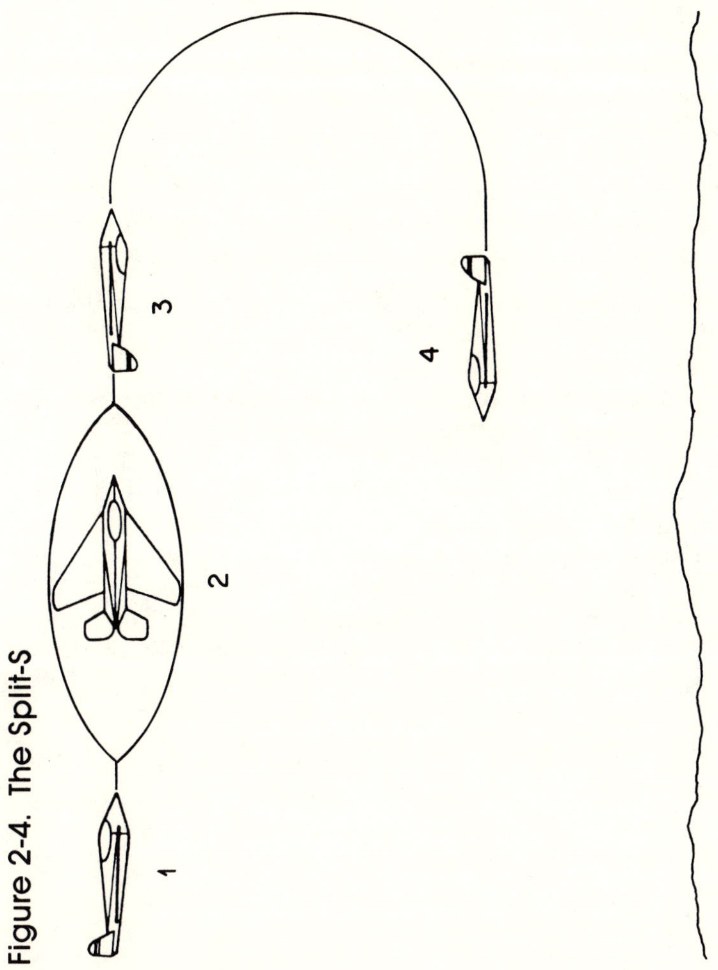

Figure 2-4. The Split-S

The Stall Turn

The stall turn is a maneuver usually associated with light, propeller-driven aircraft, but it can also be performed with jet simulations. This is a good last maneuver after a series of vertical moves. The most difficult part of this trick is entering the maneuver flying perfectly vertically. If your simulation program offers views out the side of the cockpit, you can use them to position yourself so you are flying straight up.

- Fly straight and level at a speed no less than 500 knots. Pull back hard on the stick until you're flying straight up. (1)
- Continue to hold the climb as your speed decreases. (2)
- Cut your power to idle. (3)
- When the aircraft reaches stall speed, a warning should sound. Let the plane stall. The nose will come down and point at the ground. (4)
- Adjust the nose so you're pointing straight down; pick up speed, add power, and pull out of the dive. Be careful not to add too much power during the dive because you can easily exceed the *Vmax*, or maximum speed, for the aircraft. (5)

Figure 2-5. The Stall Turn

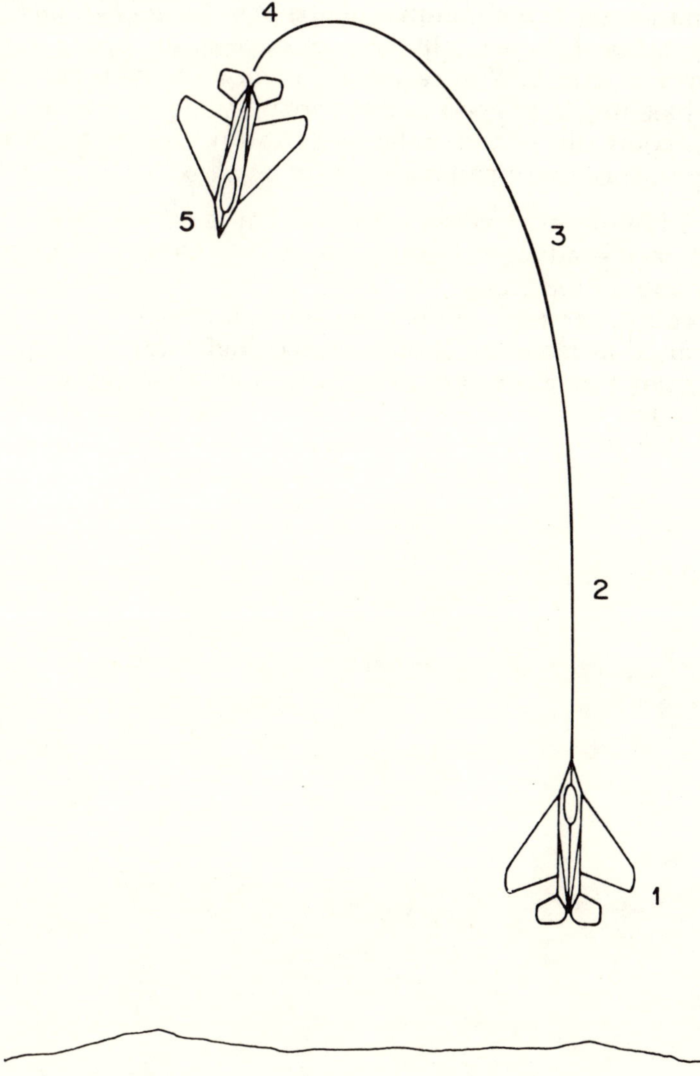

The Immelmann

This maneuver is also known as the *vertical half loop and the roll off the top.* It is basically the opposite of the split which we covered earlier. You can use this move to gain altitude while bleeding off speed and reversing your direction. You should come out of the maneuver heading 180 degrees away from your original heading.

- ⊙ Fly straight and level at a speed of at least 500 knots. (1)
- ⊙ Pull back gradually on the stick as you would when performing a loop. (2)
- ⊙ As you approach the inverted position, push forward on the stick to maintain level inverted flight. (3)
- ⊙ Perform a half aileron roll to bring the plane right side up. (4)

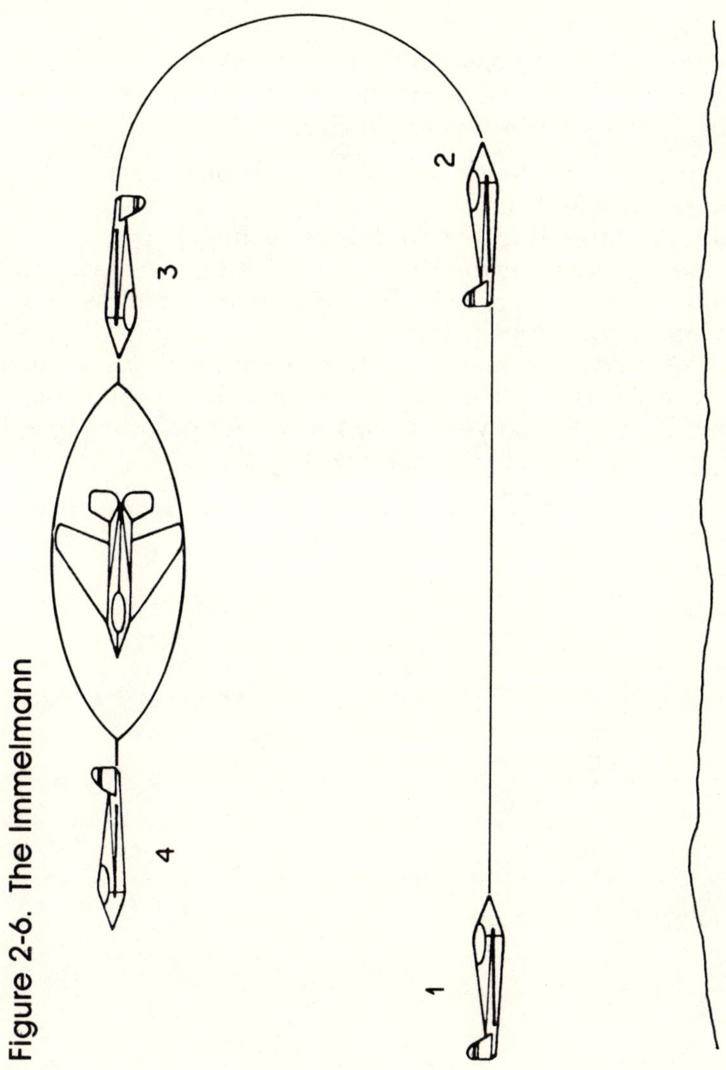

Figure 2-6. The Immelmann

The Chandelle

The chandelle looks easy at first, and it is, but it's also a precise maneuver. Not all aerobatics are hard, severe, turning moves. Some are gentle but precise.

- Fly straight and level; dive slightly to pick up a little speed and momentum. (1)
- Bank the aircraft into a 30-degree turn. (2)
- While maintaining the turn, pull back on the stick until you're in a 30-degree climb; add power as necessary to maintain your speed. (3)
- As you reach the halfway point in your 180-degree turn, start to roll out of the turn gently and bring the nose down, timing it so your wings level out on a heading 180 degrees from your original heading. (4)

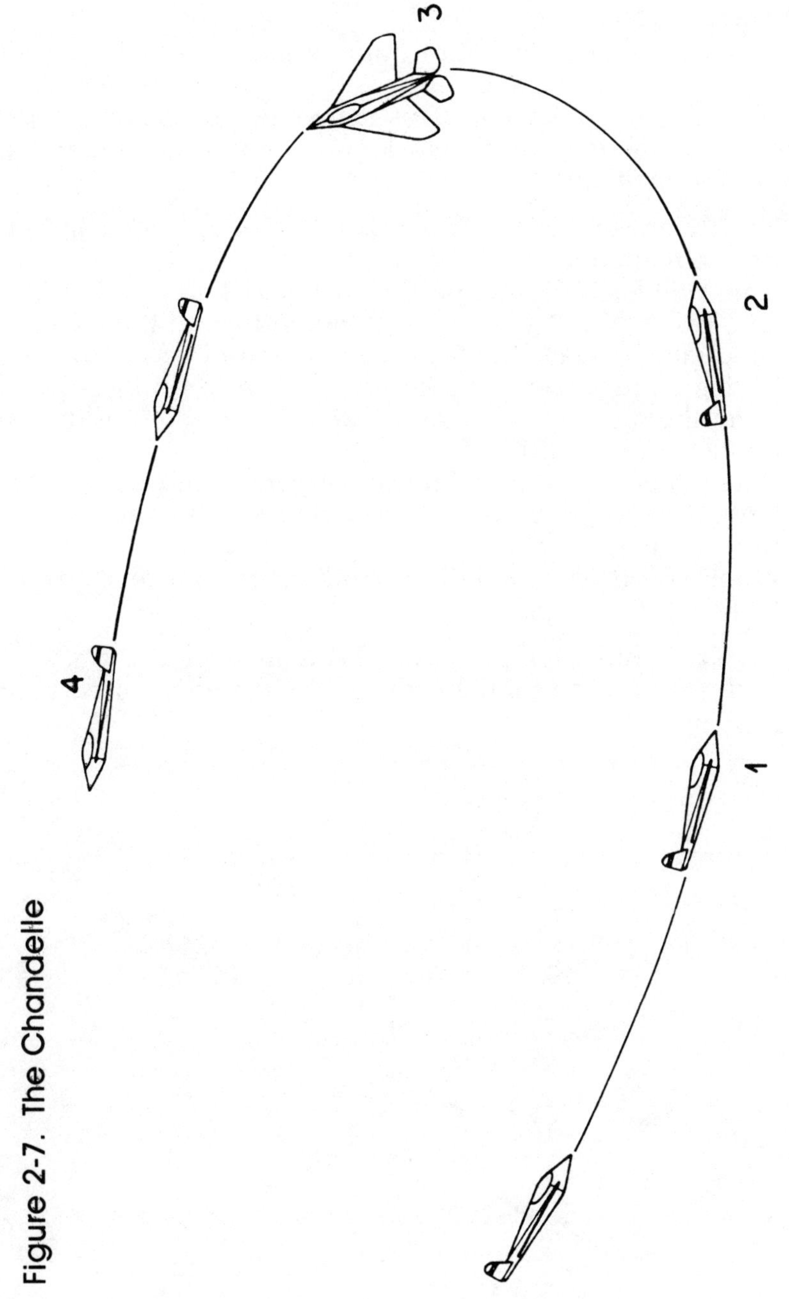

Figure 2-7. The Chandelle

The Lazy Eight

The lazy eight is an excellent training maneuver. During this maneuver you will climb, turn, and dive through an ever-changing range of speeds. This maneuver requires almost constant adjustment to the stick and attention to where you are in the maneuver.

- ⊙ Fly straight and level. Then dive to pick up speed and enter the maneuver. (1)
- ⊙ Now pull back slightly on the stick and start a roll to the left. This will put you into a climbing turn to the left. (2)
- ⊙ Gradually increase the banking angle until you're in a 90-degree turn. At this point push forward on the stick to bring the nose down to start a dive and roll out of the turn into level flight. (3)
- ⊙ When you return to the altitude at which you started, begin a climbing turn to the right, identical to your earlier turn to the left. (4)
- ⊙ When you reach a 90-degree bank, start your dive and level out the wings. (5)

You should attempt to end this maneuver at the same altitude and on the same heading as when you started your first turn.

Figure 2-8. The Lazy Eight

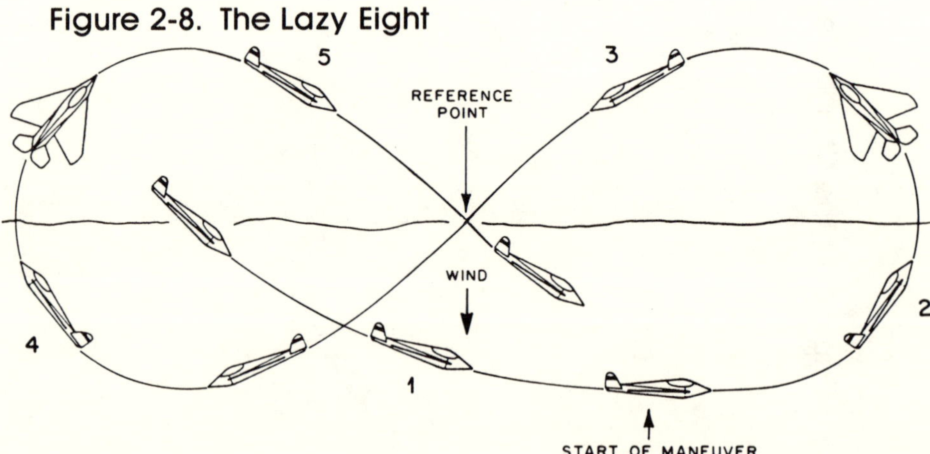

CHAPTER 3
Advanced Aerobatics

Like most complex tasks, learning aerobatics is a building process. With this in mind, take the time to master the basic maneuvers before moving on to more complex moves. Most of the advanced maneuvers in this chapter involve loops and rolls and other combinations of basic maneuvers. If you can't do them by themselves, you won't be able to put them together in combinations.

Some of these maneuvers are simply variations on moves you've already learned. Others are complex combinations of a variety of maneuvers that will take longer to master. There is no set timeframe for learning these moves, so progress at your own rate. Remember that it's better to know how to fly a few maneuvers very well than it is to fly many moves poorly. In aerobatics, as in most things, quality is more important than quantity.

In the following descriptions, where applicable, an entry speed is given. Since jet aircraft simulations cover a wide variety of aircraft with varying capabilities, the recommended airspeed is very conservative. You should be able to perform these maneuvers at much lower speeds, but when learning, it's best to use more speed than you need. Don't use too much speed, however. More speed is not necessarily better, especially with outside maneuvers.

Practice is the key. Take these moves one at a time and practice them until they're second nature. Once you start to put them together into a routine, things will happen very quickly. It's easy to become confused and disoriented. Things will also become more dangerous as you move these maneuvers close to the ground for display flying.

The Cuban Eight

The *Cuban eight* is one of the aerobatic standards. Pilots in biplanes and F-18s alike love to fly Cuban eights in demonstrations. When properly performed, this maneuver traces out a figure eight on its side. As such, it is sometimes referred to as a *horizontal eight*, but the horizontal eight is a different maneuver to be covered later. The Cuban eight consists of two modified Immelmanns performed together.

- ⊙ Fly straight and level at approximately 500 knots. Make a mental note of your altitude. (1)
- ⊙ Pull back on the stick as you would if you were performing a loop or an Immelmann. (2)
- ⊙ Continue the top of the loop on into an inverted dive. (3)
- ⊙ As the dive gets steeper, begin to roll. (4)
- ⊙ Continue to roll until you're in a level-winged 45-degree dive. Pull out of the dive at your entry altitude. (5)
- ⊙ Repeat the maneuver in the opposite direction. (6)

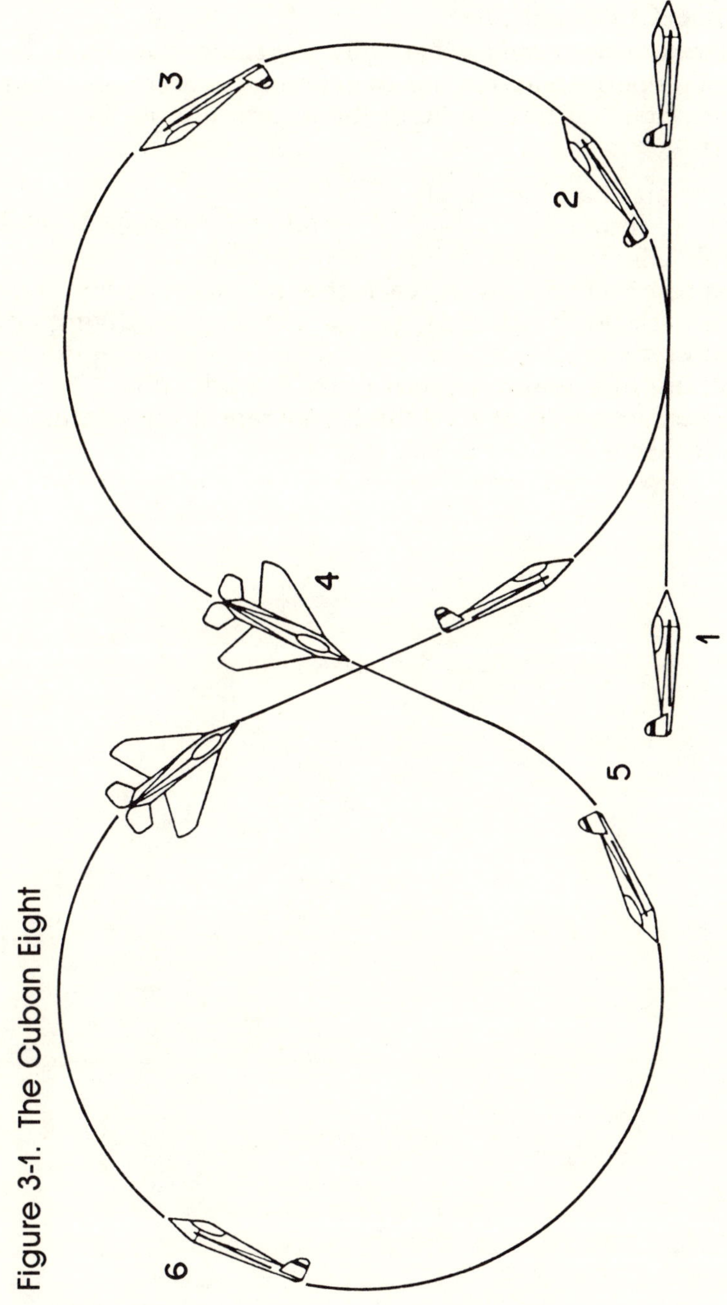

Figure 3-1. The Cuban Eight

Reverse Cuban Eight

The *reverse Cuban eight* is basically the same maneuver, but the roll is performed during the climbing portion of the maneuver. Start in the middle of the figure, not on the edge of a loop.

- ⊙ Fly straight and level. (1)
- ⊙ Pull up into a 45-degree climb and start an aileron roll while climbing. (2)
- ⊙ Stop the roll when you reach the inverted position. (3)
- ⊙ Pull back on the stick to start a dive, like the diving portion of a loop. (4)
- ⊙ Pull out of the loop at your entry altitude. (5)
- ⊙ Pull up into a 45-degree climb and repeat the process for the other half of the figure eight.

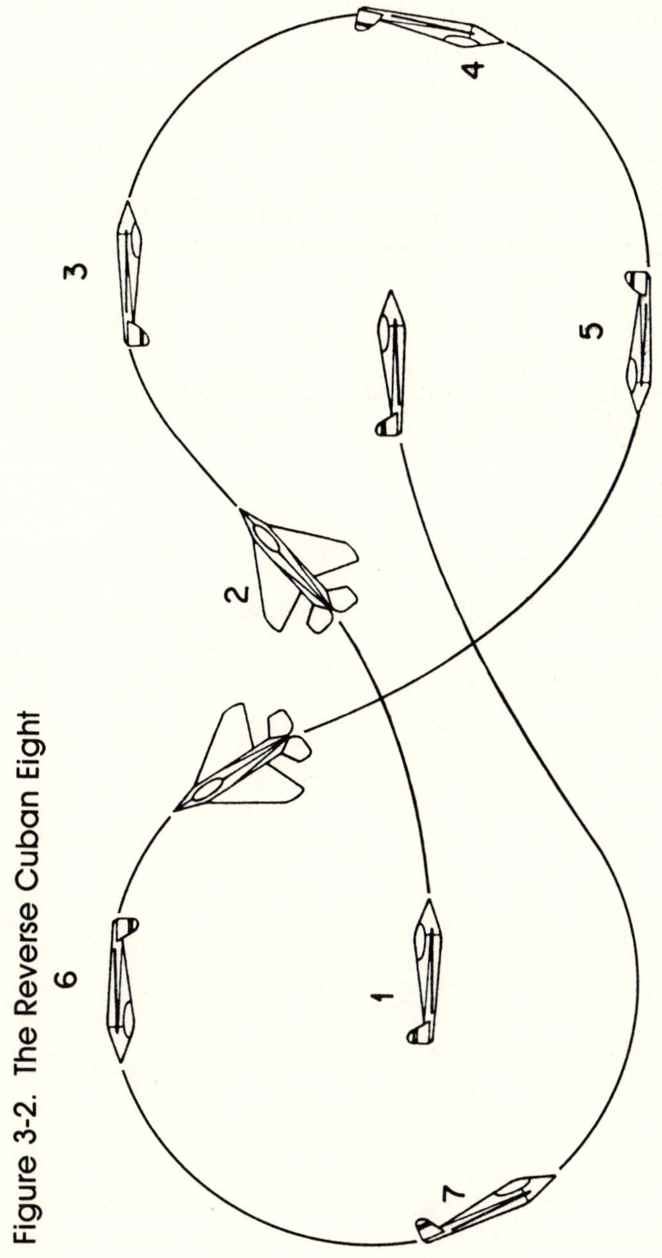

Figure 3-2. The Reverse Cuban Eight

The Outside Loop

Outside maneuvers generally cause negative G forces, meaning that instead of being pushed down into your seat, you're being pulled out of your seat. It's similar to the uncomfortable feeling you get when going over the top in a roller coaster. Fortunately, a computer pilot won't feel a thing. If your simulation takes negative G forces into account, you must perform all outside maneuvers gently to keep from losing control.

- ⊙ Fly straight and level at 600 knots or more, then roll to a level inverted position. (1)
- ⊙ Now push forward on the stick to bring the nose up, and hold it there. (2)
- ⊙ The plane will climb until you're at the top of the loop, right side up. But you're only halfway there. Here comes the scary part. (3)
- ⊙ Continue to push forward on the stick and the aircraft will start to dive. (4)
- ⊙ You should be in the bottom half of the loop. Continue to push forward on the stick until the horizon reappears and you're in level inverted flight once again.

Figure 3-3. The Outside Loop

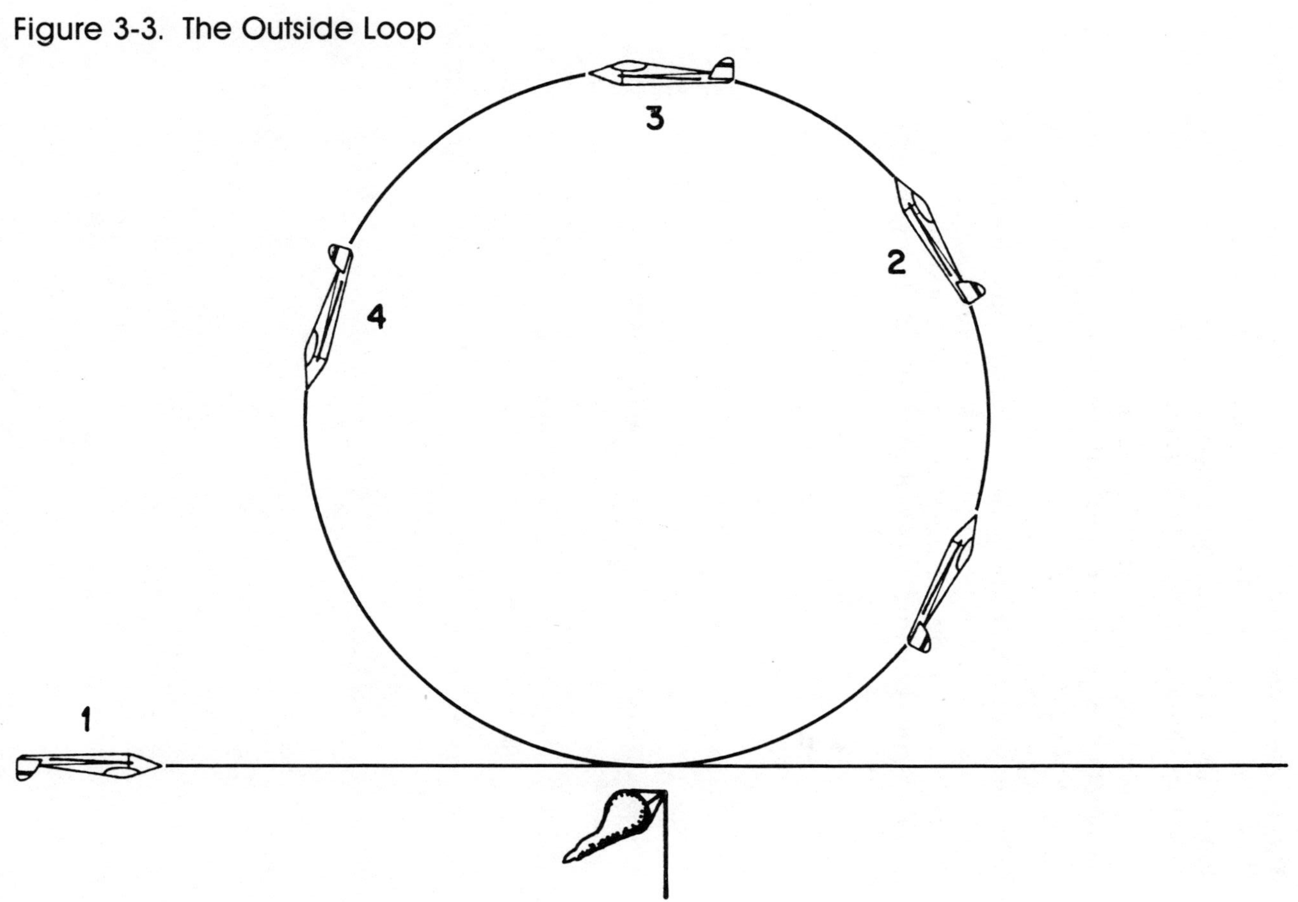

The Horizontal Eight

The *horizontal eight* is basically three-quarters of an inside loop followed by three-quarters of an outside loop. This is an excellent demonstration maneuver.

- ⊙ Fly straight and level at 500 knots or more. Pull back on the stick as if you are performing a loop. (1)
- ⊙ As you come over the top of the loop, continue into a 45-degree inverted dive. (2)
- ⊙ As you approach your entry altitude, push forward on the stick as if you are performing an outside loop. (3)
- ⊙ As you cross the top of your outside loop, continue into a 45-degree dive and pull out in level flight at your entry altitude. (4)

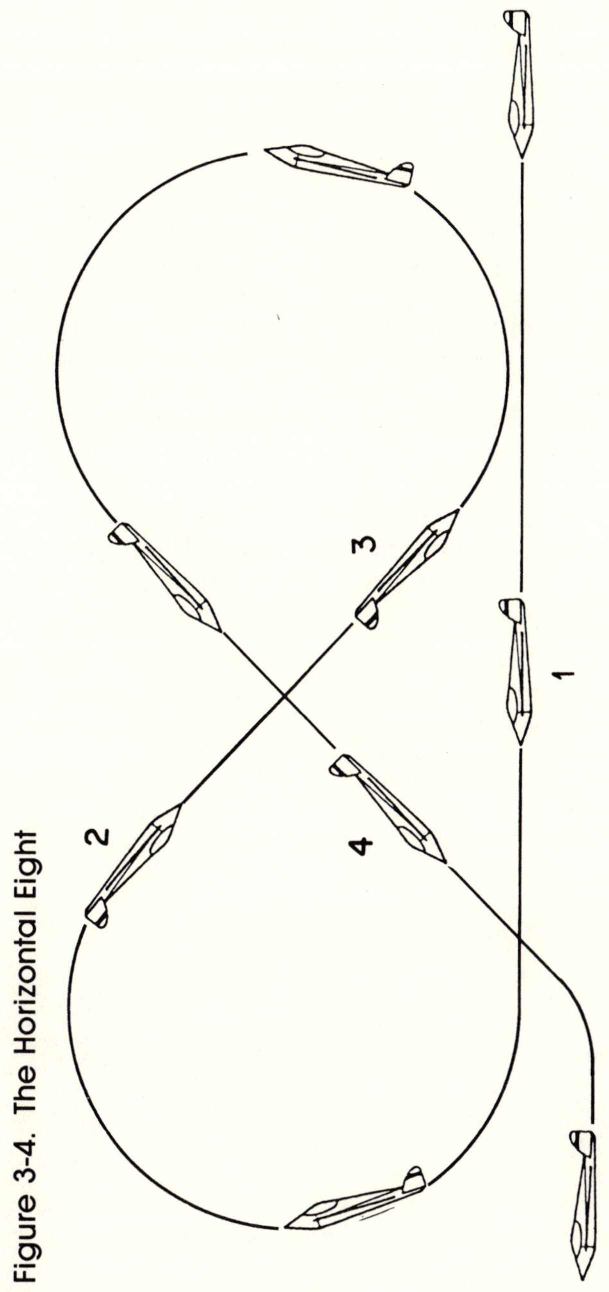

Figure 3-4. The Horizontal Eight

The Vertical Eight

The *vertical eight* is the last in the series of figure-eight maneuvers. It consists of making a transition from an inside loop to an outside loop and back again in the vertical plane. This is another showy maneuver for aircraft with excess thrust.

⊙ Fly straight and level at 600 knots or more. Depending on your aircraft, you may need to apply maximum power during the climbing portion of the maneuver. If so, remember to power back in the descending portion. (1)
⊙ Pull back on the stick as you would when entering a loop. (2)
⊙ When you reach the top of the loop, rather than going over and starting down, push forward on the stick and start an outside loop. (3)
⊙ At the bottom of your completed outside loop, pull back on the stick and complete the second half of the inside loop. (4)

Figure 3-5. The Vertical Eight

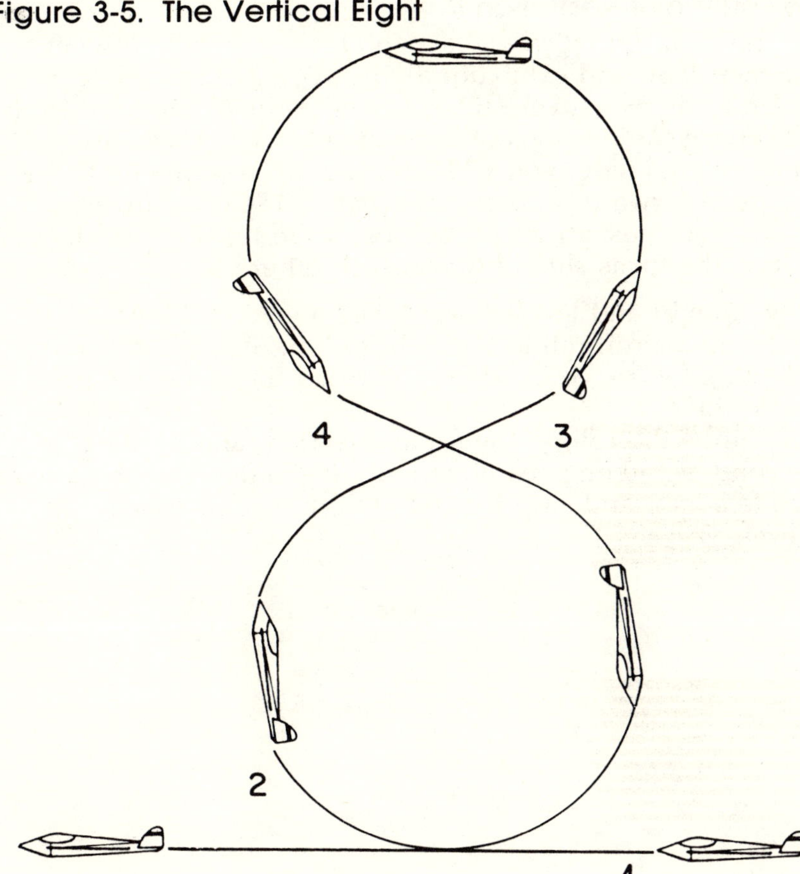

The Four-Point Hesitation Roll

The *four-point hesitation roll* is a good maneuver for low-level demonstration and exhibition flying. You'll want to begin and end this maneuver on the same heading. The problem is that during the portions of the maneuver when you are flying on your side, you will be turning. To compensate for this, push forward on the stick when flying on your side. Also, if your hesitations at each point are for the same length of time, the turns should cancel each other.

- ⊙ Fly straight and level at any speed above stall level. (1)
- ⊙ Start an aileron roll to the right or left. Stop the roll and hesitate for a second when you reach the 90-degree position. (2)
- ⊙ Continue the roll, in the same direction, and hesitate for a second at 180 degrees (inverted), at 270 degrees (the other knife edge), and stop the roll with the wings exactly level. (3)

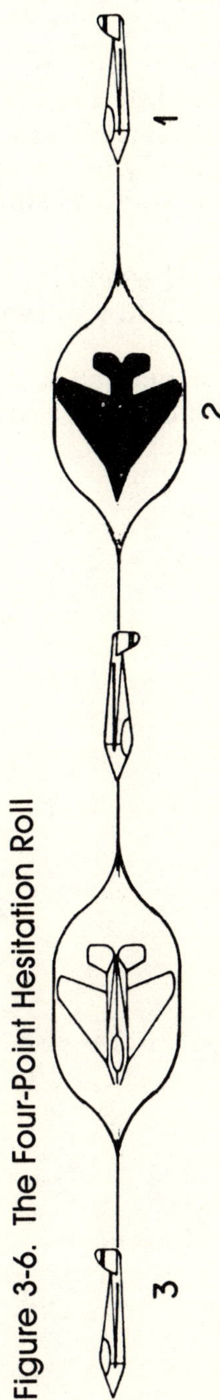

Figure 3-6. The Four-Point Hesitation Roll

The Vertical Roll

The *vertical roll* can be performed by itself or as part of any other maneuver that requires a straight up climb. The main problem with this maneuver, as with all vertical maneuvers, is knowing when you are flying perfectly straight up. Use the different view function of the simulation to help you locate the vertical.

- ⊙ Fly straight and level at 500 knots or more. (1)
- ⊙ Pull back on the stick until you're in a 90 degree (vertical) climb. (2)
- ⊙ Now apply full left or right stick and perform an aileron roll. At this point you can level off, or continue the climb, into another maneuver. (3)

Figure 3-7. The Vertical Roll

The Spin

There are a number of different kinds of spins:

- ⊙ Flat spins
- ⊙ Inverted spins
- ⊙ Power-off spins
- ⊙ Falling leaf spins

These are only a few of the spins you can perform. For your aerobatic demonstration, however, you only need one good spin. Here is how to perform the power-on spin, which is nothing more than a diving aileron roll.

- ⊙ Fly straight and level with plenty of altitude and with power at about 50 percent. (1)
- ⊙ Simultaneously, push the stick forward to start a dive, and begin an aileron roll. (2)
- ⊙ Place the aircraft in a 90-degree dive and hold it while continuing the roll. (3)
- ⊙ You can continue to spin as long as you have sufficient altitude. Be careful not to exceed the airspeed limits of the aircraft. When you're ready to pull up, stop the roll (if you're in a good spin you may have to apply stick pressure in the opposite direction from the spin), pull back on the stick, and level out the wings. (4)

Figure 3-8. The Spin

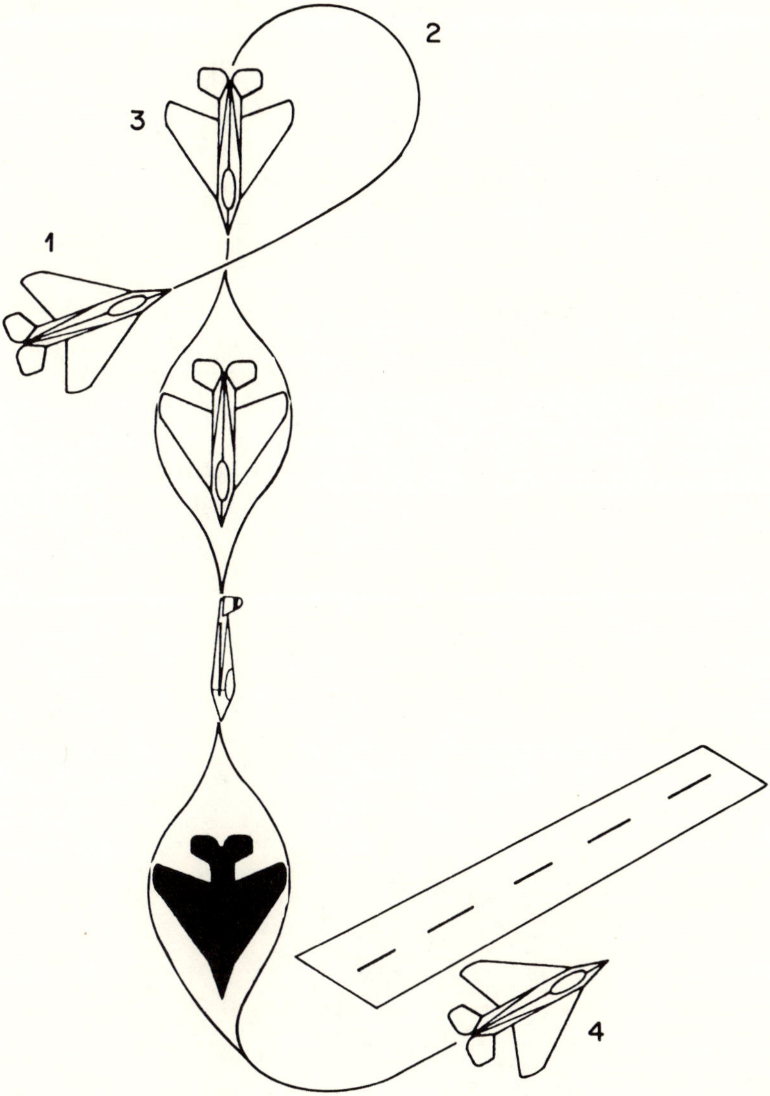

Inverted Flight and the High-Speed Low-Altitude Pass

As you may have noticed, aerobatics requires a great deal of flying upside-down. Inverted flight is not only useful in many maneuvers but can be a maneuver all in itself. Low-altitude high-speed passes are much more spectacular if performed inverted.

- ⊙ Roll to straight and level inverted flight at 400 knots or more. (1)
- ⊙ Pull back slightly on the stick to descend to an altitude of 500 feet. (2)
- ⊙ When level at 500 feet, perform a series of up and down *jinks*. Jinks are rapid changes of direction for a brief period of time. Remember when you are inverted you must push forward on the stick to push your nose up. (3)
- ⊙ Make a 180-degree turn and level out, inverted for a second pass. Push forward on the stick to gain altitude. When you're back to at least 2000 feet, roll right side up. (4)

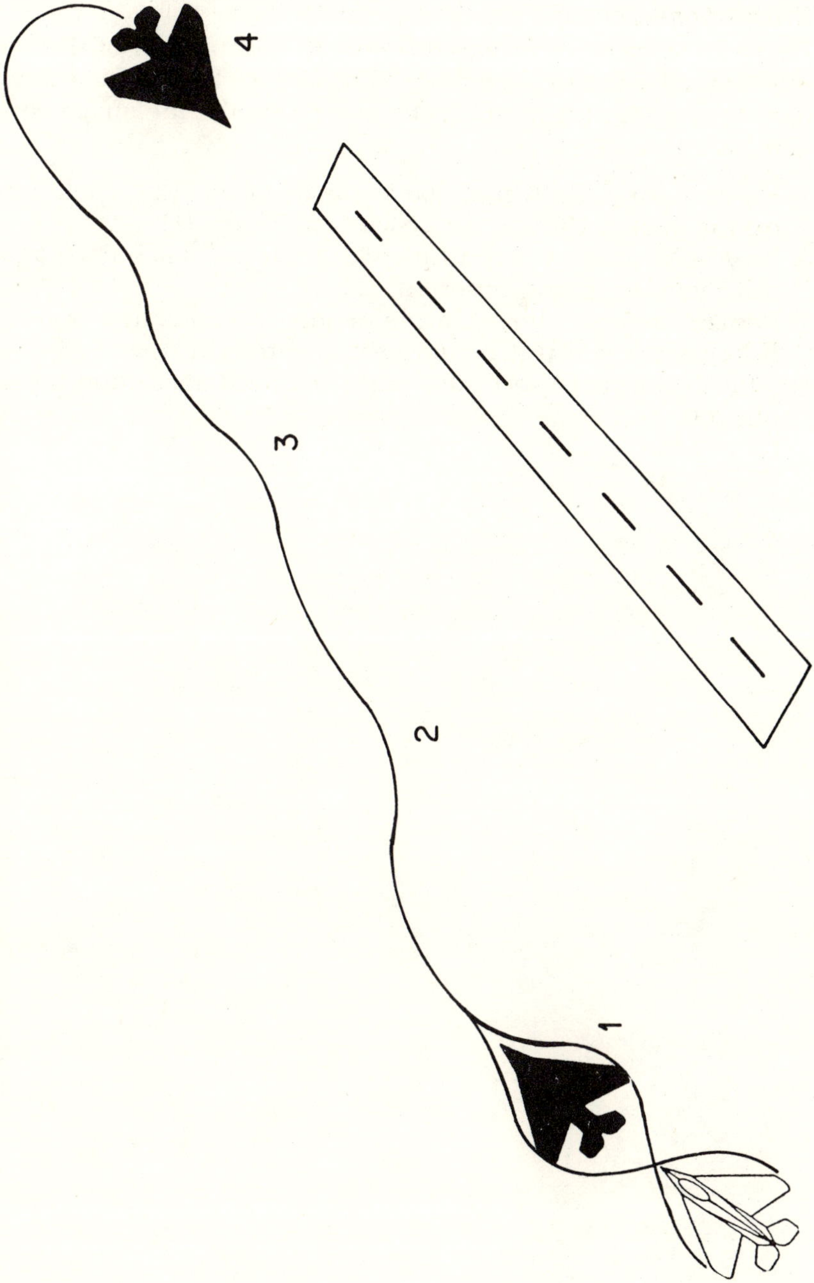

Figure 3-9. Inverted Flight and the High-Speed Low-Altitude Pass

The Square Loop

There are a number of variations to the loop. One of the most popular is the *square loop*. The square loop is basically a box; you must pull hard on the stick to make the sharp corners.

- ⊙ Fly straight and level at 500 knots or faster. Pull back hard on the stick until you are in vertical flight. (1)
- ⊙ Stay vertical for a second or two; then pull back hard again until you are flying inverted. (2)
- ⊙ Remain inverted for a couple of seconds and then pull back hard again until you're diving straight down. (3)
- ⊙ After a couple of seconds, pull back hard again and level out. (4)

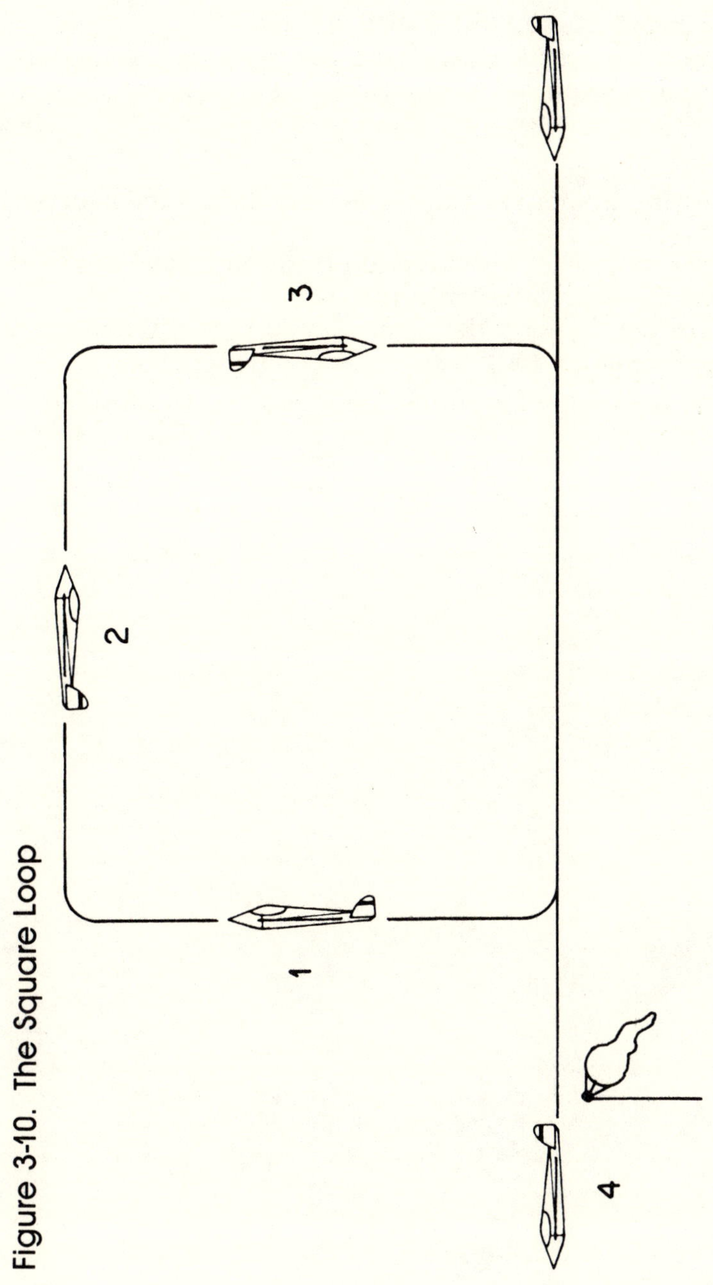

Figure 3-10. The Square Loop

The Square Loop with a Roll

The *square loop with a roll* is another variation on the loop. Square loops can have one, two, three, or four rolls (one for each side of the box), but the one-roll variety is described below.

⊙ Start the maneuver exactly as you started the square roll. (1)

⊙ When you reach the inverted position on the top of the loop, perform an aileron roll. (2)

⊙ Continue the loop the same as you would when performing the square loop. (3)

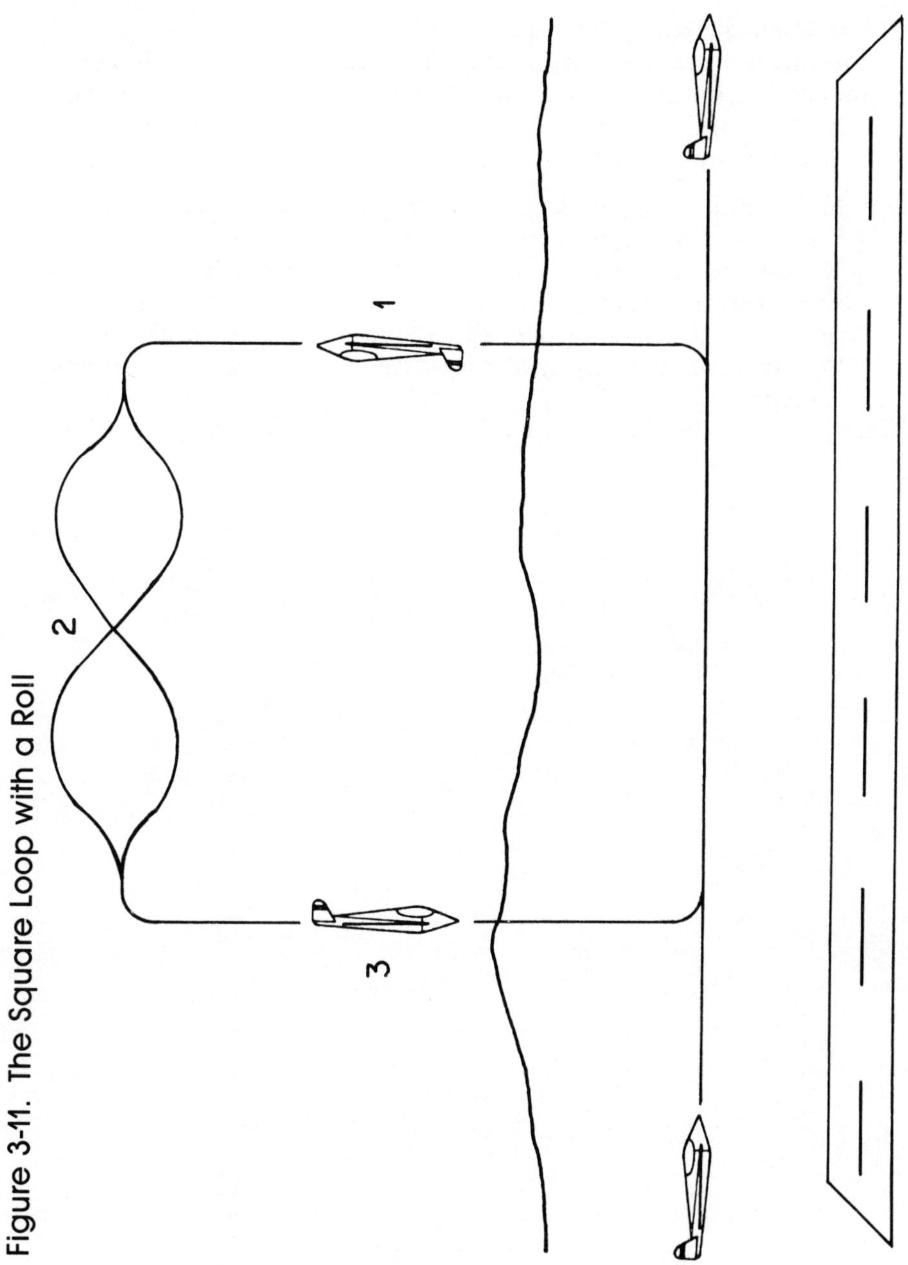

Figure 3-11. The Square Loop with a Roll

The Outside Turn

All simulations won't allow you to perform the outside turn, but most will. The *outside turn* is a 360-degree turn with the cockpit on the outside of the turn as opposed to the normal turn where the cockpit would be on the inside of the turn.

- ⊙ Fly straight and level at 400 knots or more. Bank the aircraft into a 90-degree turn. (1)
- ⊙ Notice the direction of your heading change. Are the numbers getting larger or smaller? Now, with the aircraft flying perfectly sideways, push all the way forward on the stick. Watch your heading indicator; the numbers should reverse direction. (2)
- ⊙ Continue the turn until you have turned 360 degrees; then level off. (3)

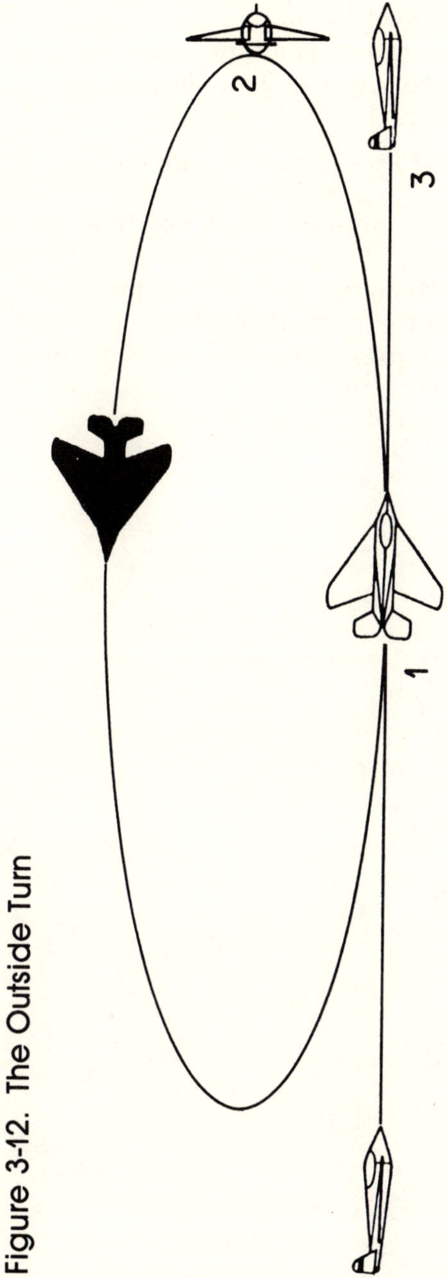

Figure 3-12. The Outside Turn

The Tuck Under Roll Turn

The tuck under roll turn is a fancy turn jet pilots sometimes perform during exhibitions. They love to use this maneuver after a low pass in front of the crowd.

- ⊙ Fly straight and level past your imaginary reviewing area. (1)
- ⊙ Pull the nose up slightly and apply hard left stick to start an aileron roll. (2)
- ⊙ When you have completed three-quarters of the roll (on your side with the cockpit facing to the right), stop the roll and pull back hard on the stick to perform a maximum rate turn to the right. (3)

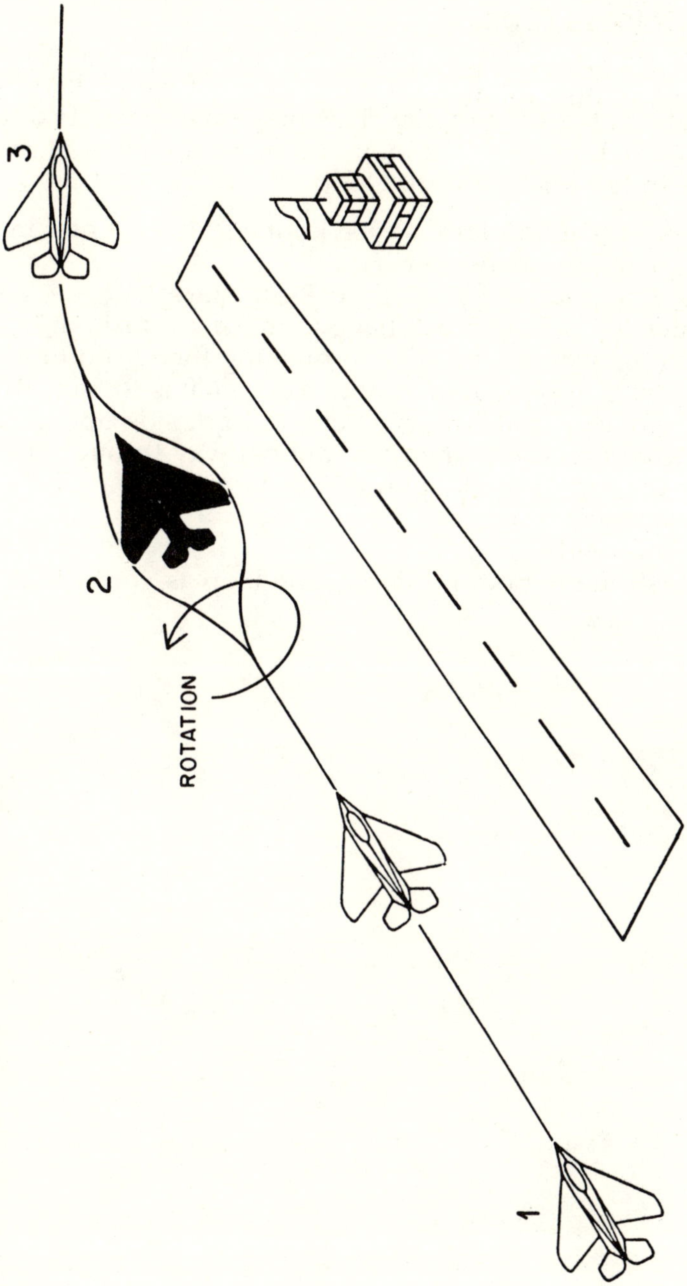

Figure 3-13. The Tuck Under Roll Turn

The Chinese Loop

There are a number of variations on the loop that involve rolls at various points in the maneuver. One of the more popular is the *Chinese loop.* This maneuver consists of the bottom half of a normal loop and one complete roll that takes place in the top half of the loop.

- Fly straight and level at 500 knots or more. Enter into a normal loop maneuver. (1)
- As you reach the vertical position, start your roll. As you enter the roll, you will have to make a transition from pulling back on the stick to pushing forward on the stick in order to remain in a loop. Your goal is to be halfway through the roll (wings level and right side up, since you started inverted) when you are halfway through the loop (at the top of the loop in the same position as you would be during an outside loop). (2)
- Continue the loop. (3)
- Finish the bottom portion of the loop as you would normally. (4)

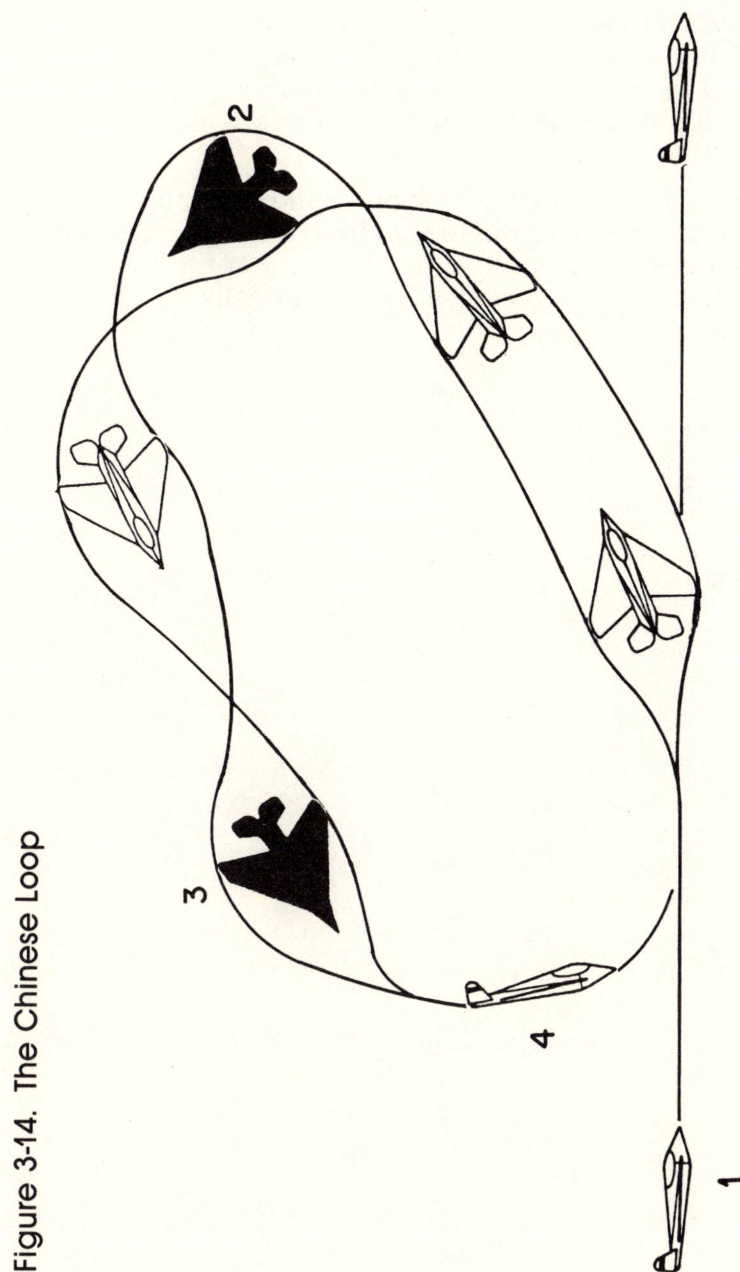

Figure 3-14. The Chinese Loop

The Avalanche

Another variation on the loop involving a roll is the *avalanche*. This maneuver consists of a normal loop with a roll on the top. It can also be performed as a *double avalanche* with two rolls on the top of the loop.

- ⊙ Enter a loop maneuver as you would normally. (1)
- ⊙ As you come over the top, in inverted flight, perform a quick aileron roll. (2)
- ⊙ Complete the loop as you would normally. (3)

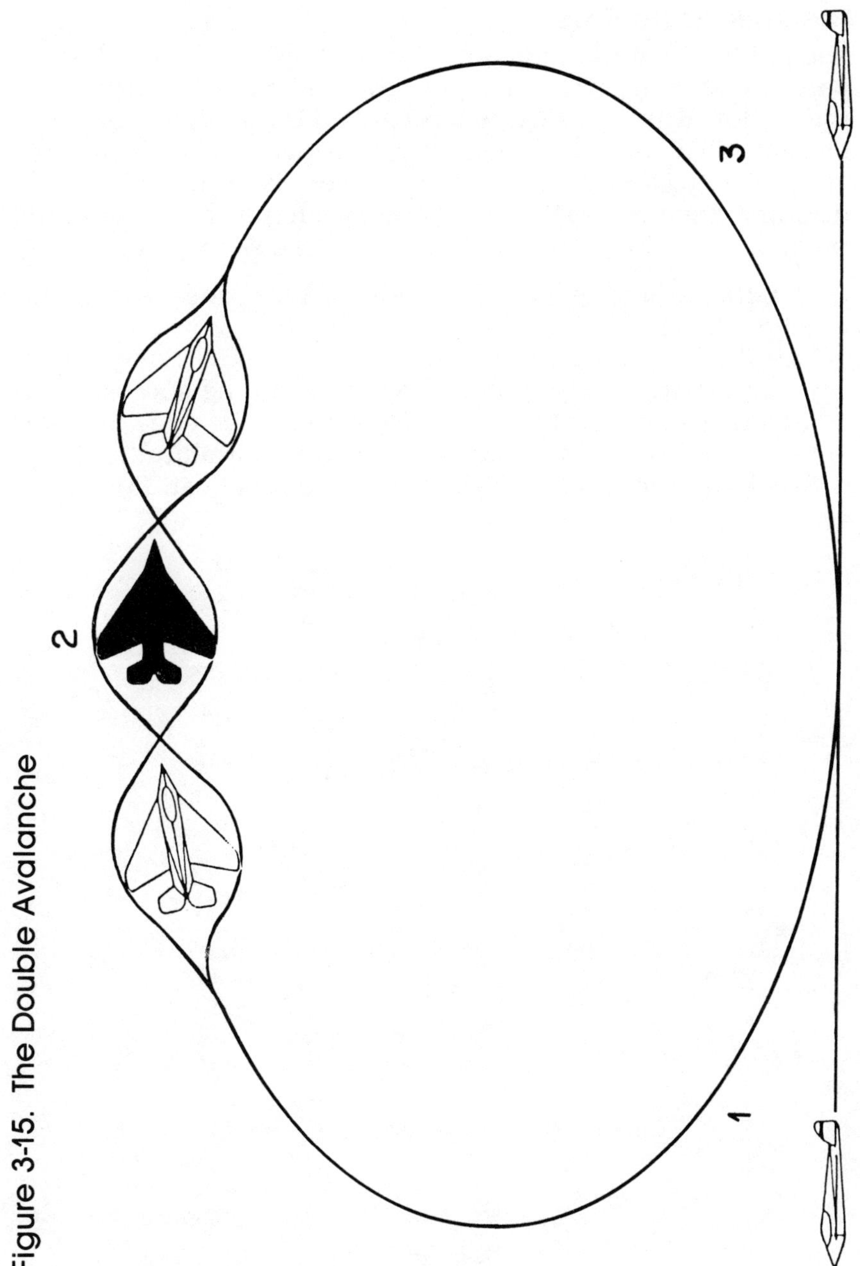

Figure 3-15. The Double Avalanche

The Knife-Edge Pass

A good low-altitude maneuver is a *knife-edge pass*. The hard part of this maneuver is keeping the plane in a straight line; the aircraft wants to turn when banked on its side. This can be controlled somewhat by applying forward stick once you are perfectly sideways. Another problem is losing altitude during a knife edge. You can compensate for this by entering the maneuver in an attitude with the nose slightly high.

- ⊙ Fly straight and level at 300 knots or more and descend to 500 feet. (1)
- ⊙ Pull the nose up slightly and roll until you're flying perfectly sideways. Use forward pressure on the stick to maintain a constant heading. Hold this knife-edge position for several seconds. Keep your eye on your altitude. (2)
- ⊙ Roll back into level flight and climb to 2000 feet. (3)

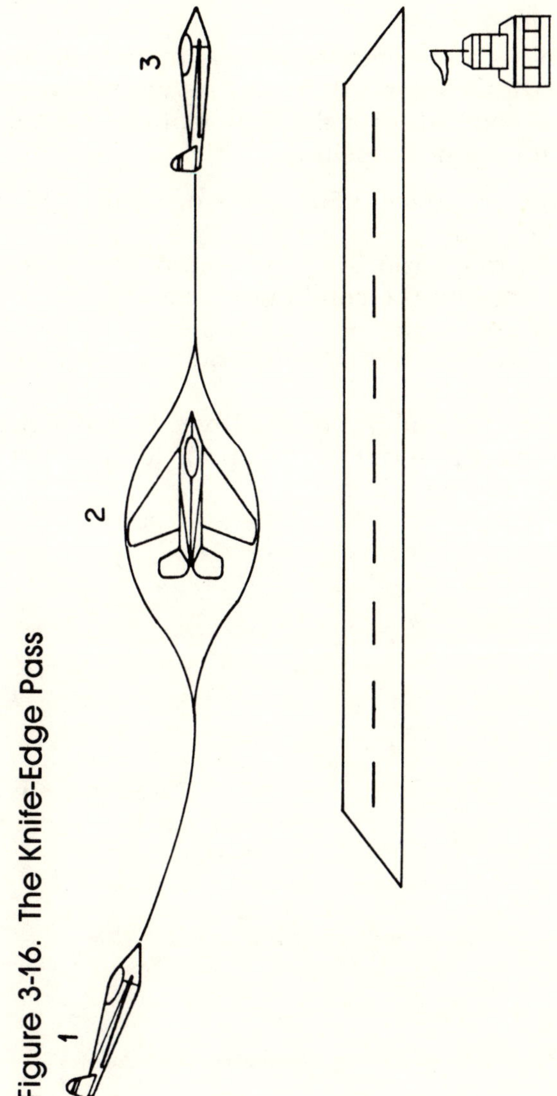

Figure 3-16. The Knife-Edge Pass

The Totem Pole

The *totem pole* is a spectacular vertical maneuver that combines the vertical roll and the spin. As with all vertical maneuvers, use your left or right view feature, if available, to help locate the vertical position.

- ⊙ Fly straight and level at 500 knots or more. Pull up into a vertical climb. (1)
- ⊙ Perform a vertical roll. At your option, several rolls can be performed during the climbing phase. (2)
- ⊙ Cut power and perform a stall turn. (3)
- ⊙ With the nose pointed straight down, perform a series of spins. (4)
- ⊙ Add power and pull out of the dive. The maneuver looks best when you pull out in the opposite direction from the direction you entered. (5)

Figure 3-17. The Totem Pole

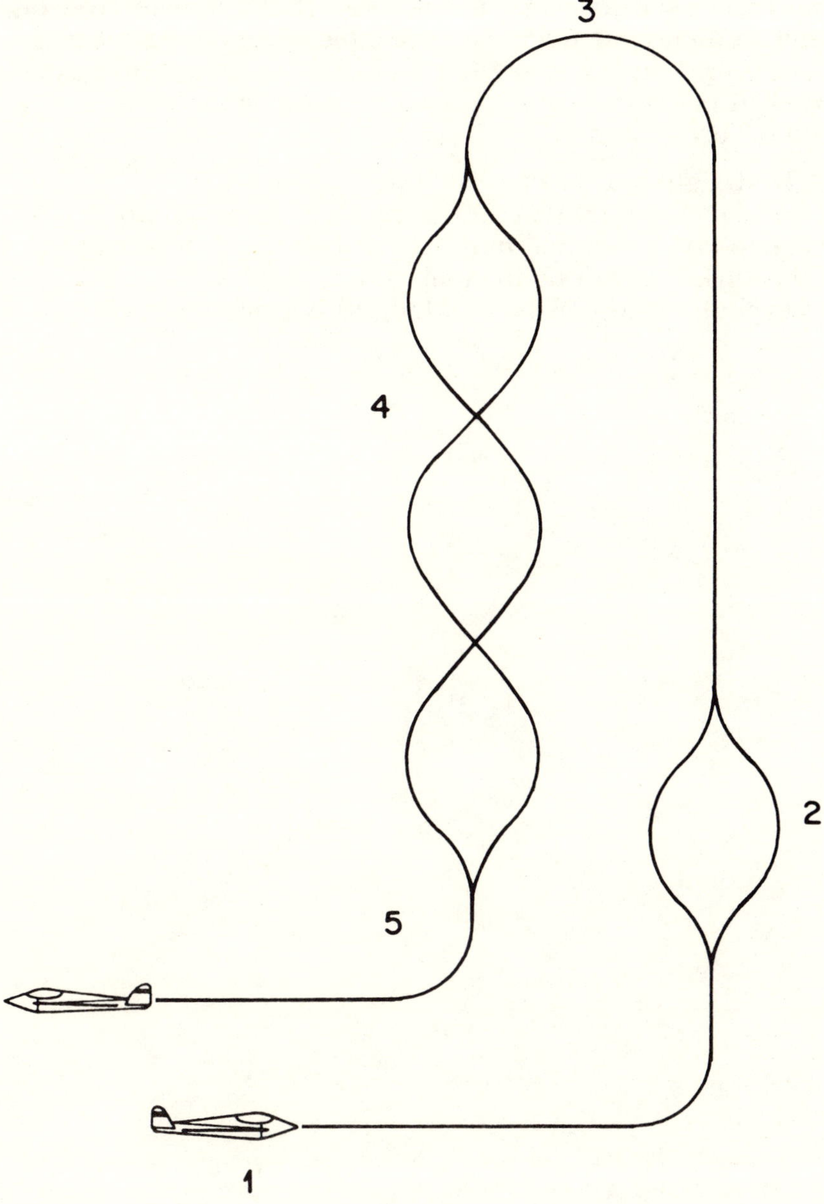

The Climbing S

This maneuver can be performed as a climbing maneuver or a descending maneuver. This example covers the climbing maneuver. You can also enter and exit this maneuver in the inverted position. It is often used as a transition maneuver to gain or lose altitude.

- ⊙ Fly straight and level at 500 knots or more. (1)
- ⊙ Pull up as though you are going to perform a loop. (2)
- ⊙ When you reach the inverted position, push forward on the stick to continue to climb. (3)
- ⊙ Level off on top of the half loop in normal flight. (4)

Figure 3-18. The Climbing S

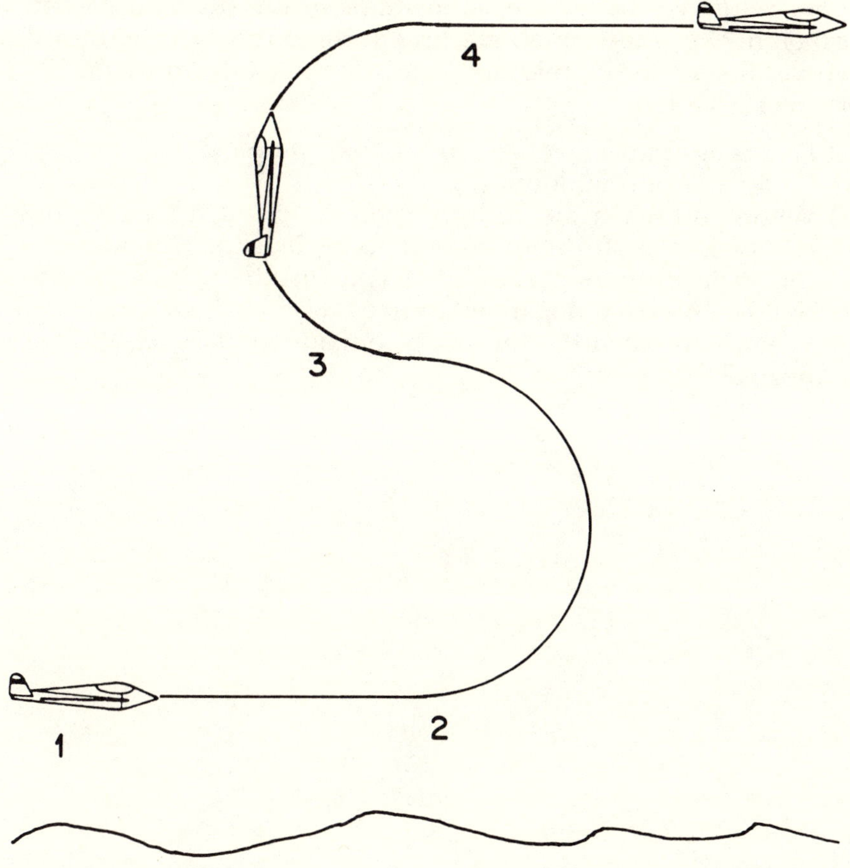

The Vertical Eight with a Middle Entry

This maneuver traces the same shape in the sky as the vertical eight, but rather than starting at the bottom of the figure, you will start in the middle. You can also perform this maneuver inverted.

- ⊙ Fly straight and level at 500 knots or more. (1)
- ⊙ Perform a normal loop. (2)
- ⊙ When you reach the bottom of the loop, push forward on the stick to perform an outside loop. Be careful to apply the stick pressure gradually or you run the risk of *red-out* due to excessive negative G forces. (3)
- ⊙ When you reach the top of the outside loop, level off in normal flight. (4)

Figure 3-19. The Vertical Eight with a Middle Entry

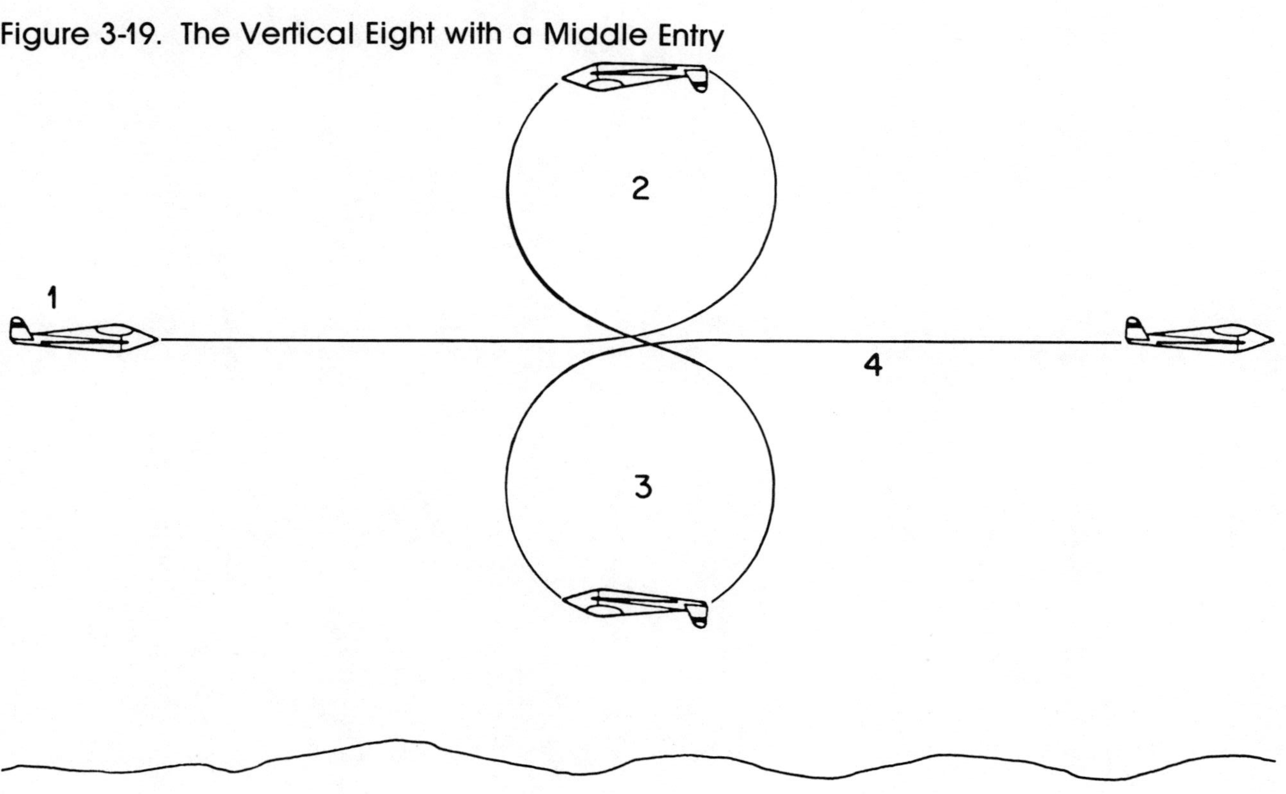

CHAPTER 4
Sequence Flying

Putting together several maneuvers is called *sequence flying.* But this broad term includes many other smaller categories of flying, including stunt flying, competitive aerobatics, and display flying. They all involve flying a number of maneuvers in a sequence, but have different results.

For the purposes of this book, competitive aerobatic flying and display flying are gathered into one discussion. This section will focus on learning sequences of maneuvers designed to improve your flying skills and, more important, showing your flying skills and the capabilities of your computer and software.

The language of sequence flying is called *Aresti.* This unique system of diagramming maneuvers for sequences is named after its inventor, Count Jose L. Aresti of Spain. He devised the system to help keep track of maneuvers in international aerobatic competition.

In Aresti, each maneuver has its own symbol. By combining these symbols, you can form logical sequences. The basic building block symbols of the Aresti system are shown in Figure 4-1.

Figure 4-1. The Symbols of Aresti

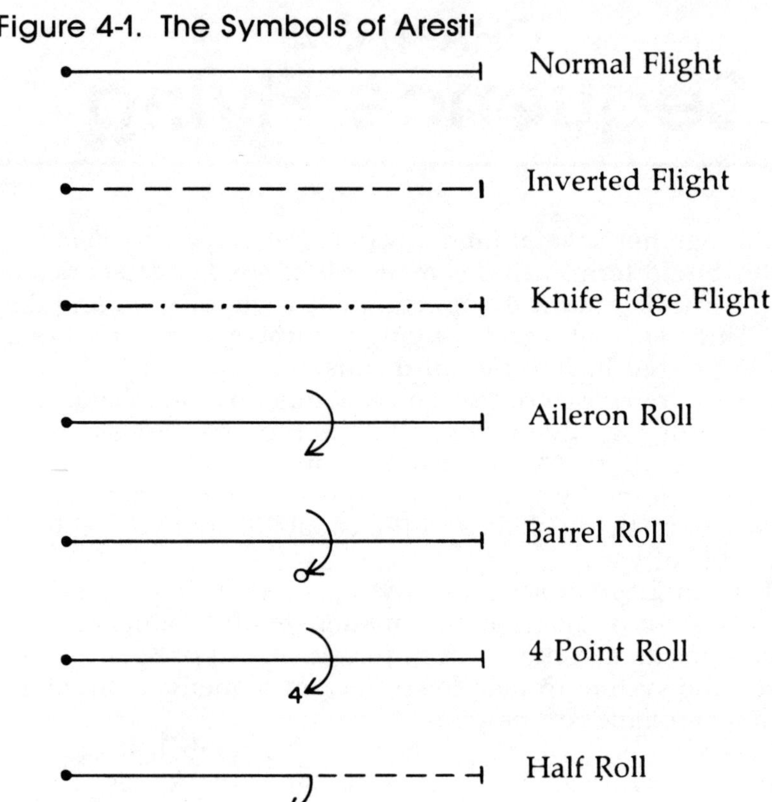

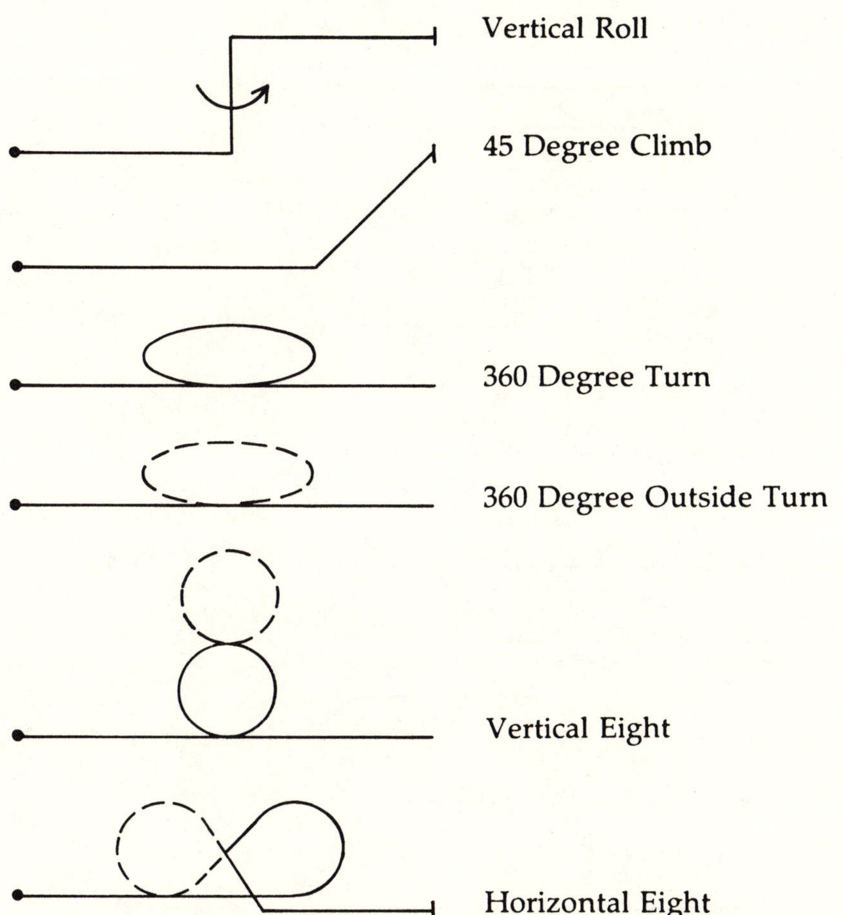
Vertical Roll
45 Degree Climb
360 Degree Turn
360 Degree Outside Turn
Vertical Eight
Horizontal Eight

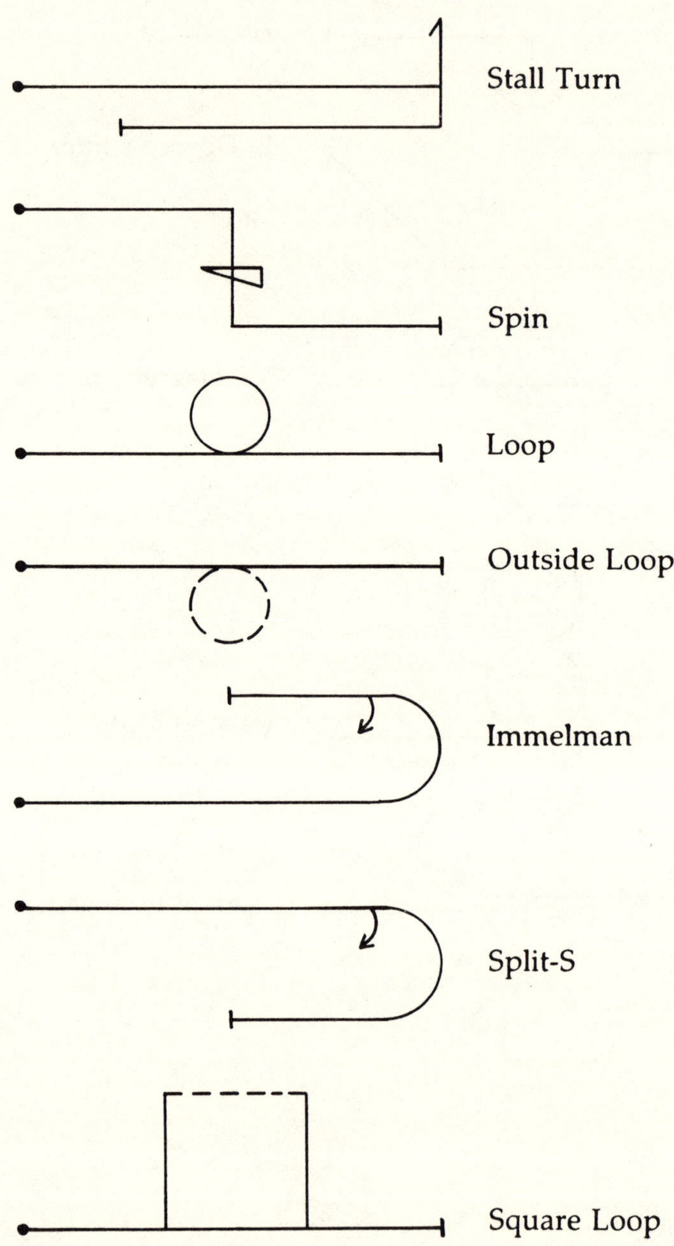
Stall Turn
Spin
Loop
Outside Loop
Immelman
Split-S
Square Loop

These symbols may look complicated at first, but a careful inspection reveals that they are very descriptive of the maneuvers they represent. By combining these symbols, you can manufacture thousands of multipart maneuvers, and by combining maneuvers you can form air show type sequences.

Learning a new routine is best handled in small chunks. Select three or four maneuvers and practice them together. When they're familiar to you, move on to the next group of moves. Once you're familiar with the whole routine in small groups of maneuvers, you can put them together in a sequence.

Basic Routines

A good sequence is more than the sum of its individual parts. A good sequence depends on how these parts fit together. Transition is very important. Moving smoothly from one maneuver to another is the mark of an accomplished aerobatic pilot. The following basic routines are good training tools to practice your transitions and location. Try to stay over a landmark on the ground to simulate the audience location. Note that these sequences incorporate frequent direction changes to keep the plane near the crowd.

Basic Sequence #1

Basic Sequence #1 is a descending sequence, so start at a great altitude. You won't regain the altitude lost during the two descending maneuvers.

1. Aileron roll
2. Split-S
3. Barrel roll
4. Spin
5. Loop
6. A 180-degree turn
7. Half roll to inverted position and half roll back
8. A 180-degree turn
9. Dive and low-level pass

Figure 4-2. Basic Sequence #1

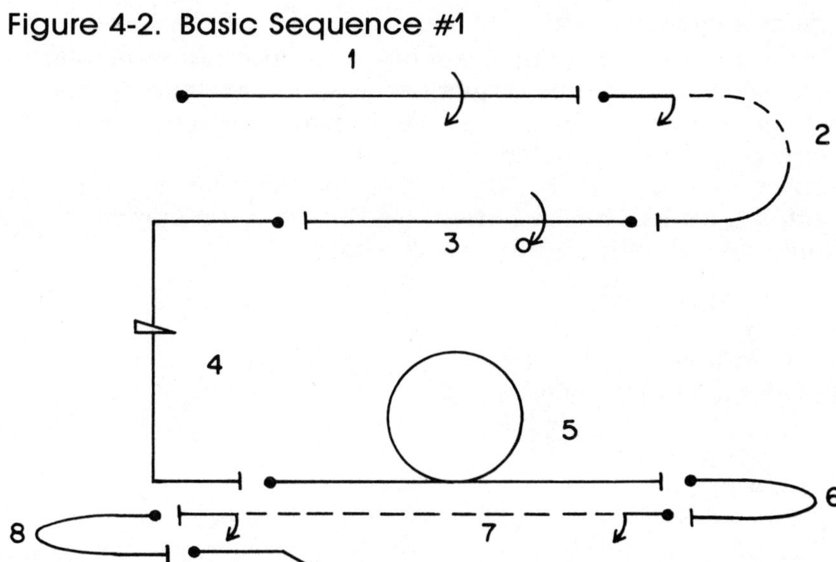

Basic Sequence #2

This sequence uses both ascending and descending maneuvers. Most competitive sequences use several climbing maneuvers because you can only do so many descending maneuvers before you run out of altitude. The dotted line between maneuvers 4 and 5 serves only to connect them. Dotted lines like this are often used to keep the maneuvers from being drawn on top of each other.

1. A ¾ square loop
2. Spin
3. A 1¼ loop
4. Double vertical roll
5. Split-S
6. Cuban eight
7. Stall turn
8. Aileron roll

Figure 4-3. Basic Sequence #2

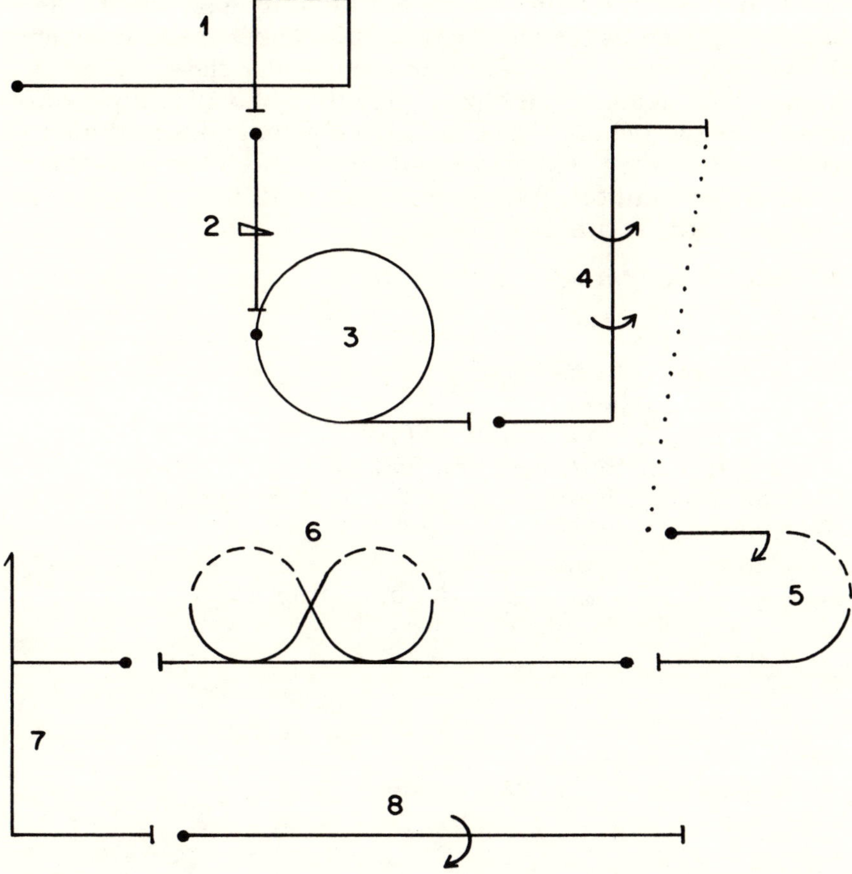

Intermediate Routine

In competitive aerobatics each pilot must fly a series of maneuvers known as *compulsory sequences*. These maneuvers are the same for every competitor in a particular class. After the compulsory sequence, the pilot may fly a short freestyle routine of his own choosing. The following sequence is similar to the compulsory sequence for flyers in the Intermediate Class of competition. As you move up in difficulty level, more maneuvers are included.

1. Aileron roll on a 45-degree climb
2. A $1\frac{1}{2}$ spin
3. Square loop
4. A $\frac{5}{8}$ loop with half roll
5. Half square loop
6. A 90-degree dive with a half roll
7. Stall turn with descending half roll
8. Half loop, half roll
9. Aileron roll
10. Half roll, half loop
11. Half roll on a 45-degree climb, quarter loop
12. Double aileron roll

Figure 4-4. Intermediate Sequence

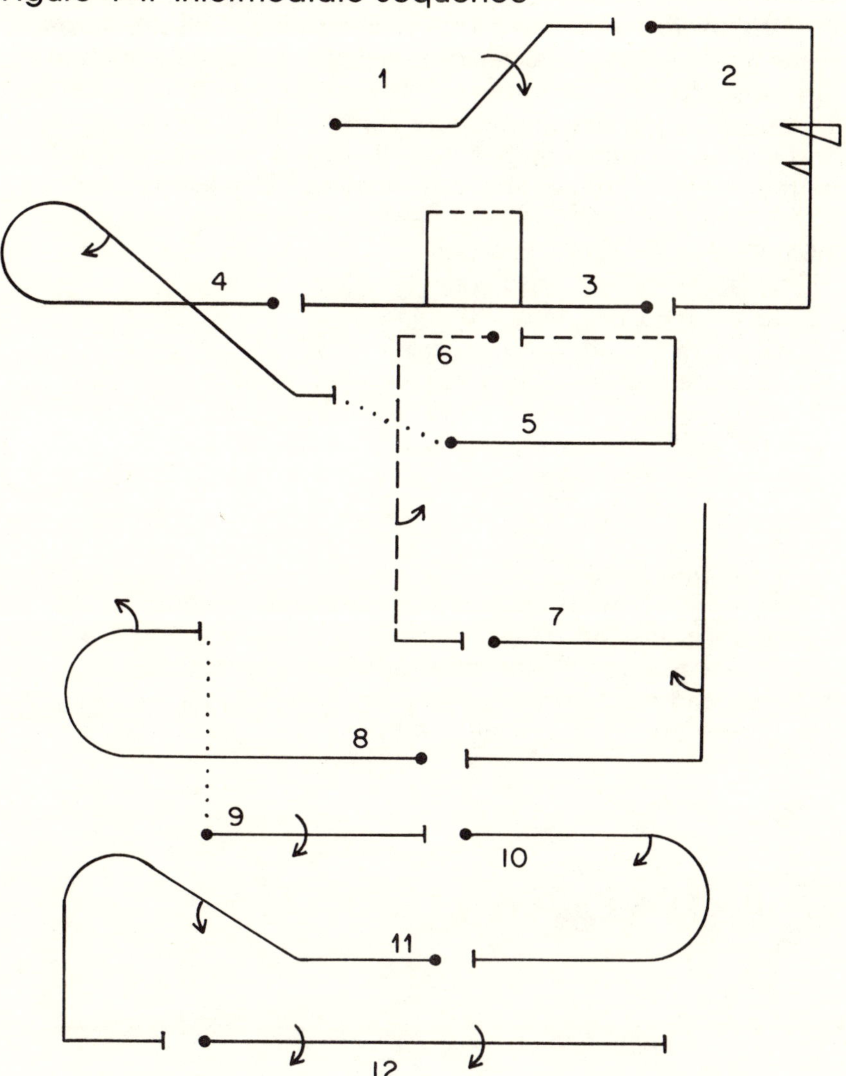

Advanced Routines

Moving on to advanced routines, we pull out all the stops. Here you will find a larger number of maneuvers, fractions of loops and rolls, and a great deal of outside flying. At the advanced stage, the maneuvers come rapidly, one after the other, so it becomes more important than ever to exit each maneuver properly so you can move directly to the next.

Advanced Sequence #1

1. 45-degree climb with roll
2. 45-degree dive with roll
3. Climb with half loop, inverted pull out
4. Outside half loop, normal flight pull out
5. Square loop with roll during climbing leg
6. Half roll to inverted position and outside loop
7. Half roll, half outside loop with inverted pull out
8. Horizontal eight from inverted entry
9. Half loop
10. Four-point roll
11. Totem pole with extra half roll during descending leg (to reverse direction)
12. 360-degree maximum rate horizontal turn
13. Descending double roll, half loop, roll
14. Power dive and low-altitude pass

Figure 4-5. Advanced Sequence #1

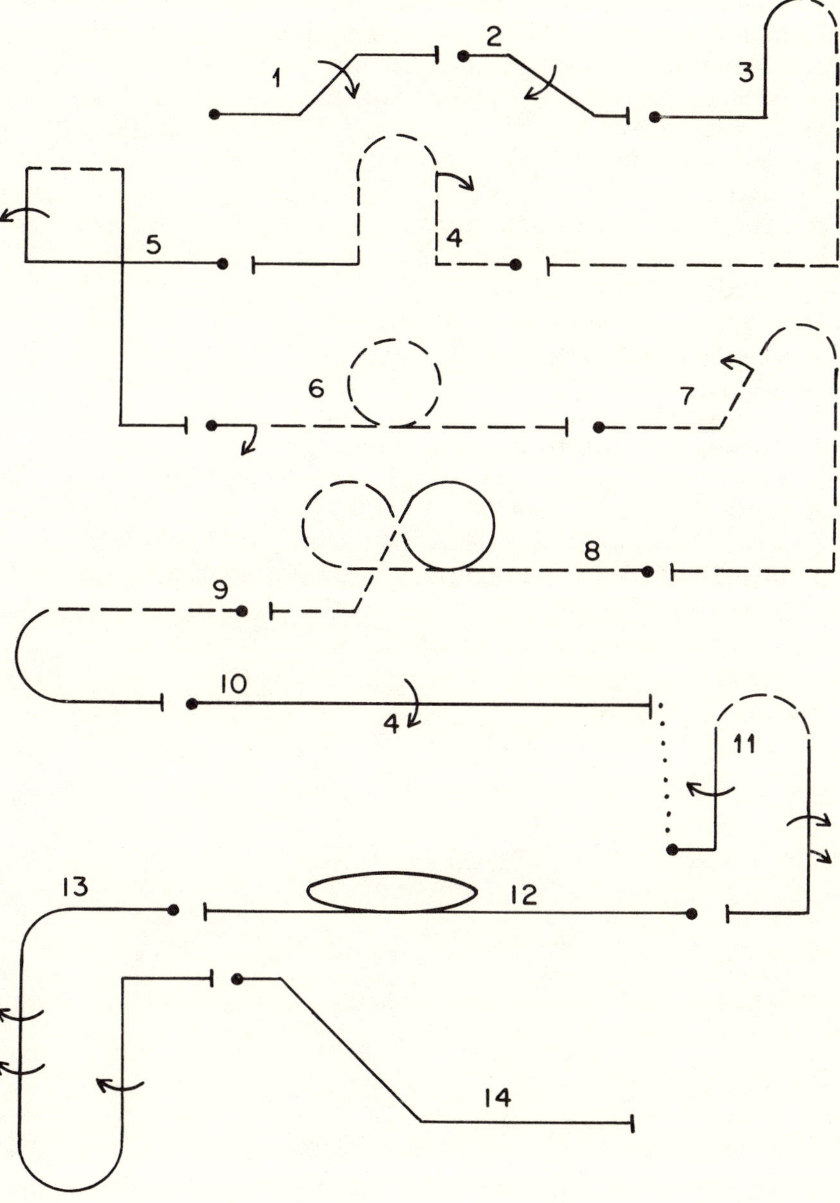

Advanced Sequence #2

1. Climbing S
2. Half roll
3. 90-degree dive with two rolls, inverted recovery
4. Inverted entry into square loop, half roll on the top
5. Vertical roll
6. 45-degree dive
7. Two loops
8. 180-degree turn, maximum rate
9. Dive to low-altitude knife-edge pass
10. 180-degree turn
11. 360-degree maximum rate turn
12. 45-degree climb with roll
13. Slip-S
14. Barrel roll
15. Tuck-under-roll turn
16. Light the after-burners for a low-altitude, high-speed pass.
17. Finish with a 90-degree ballistic climb to 30,000 feet, or until stall speed is reached if less than 30,000.

Figure 4-6. Advanced Sequence #2

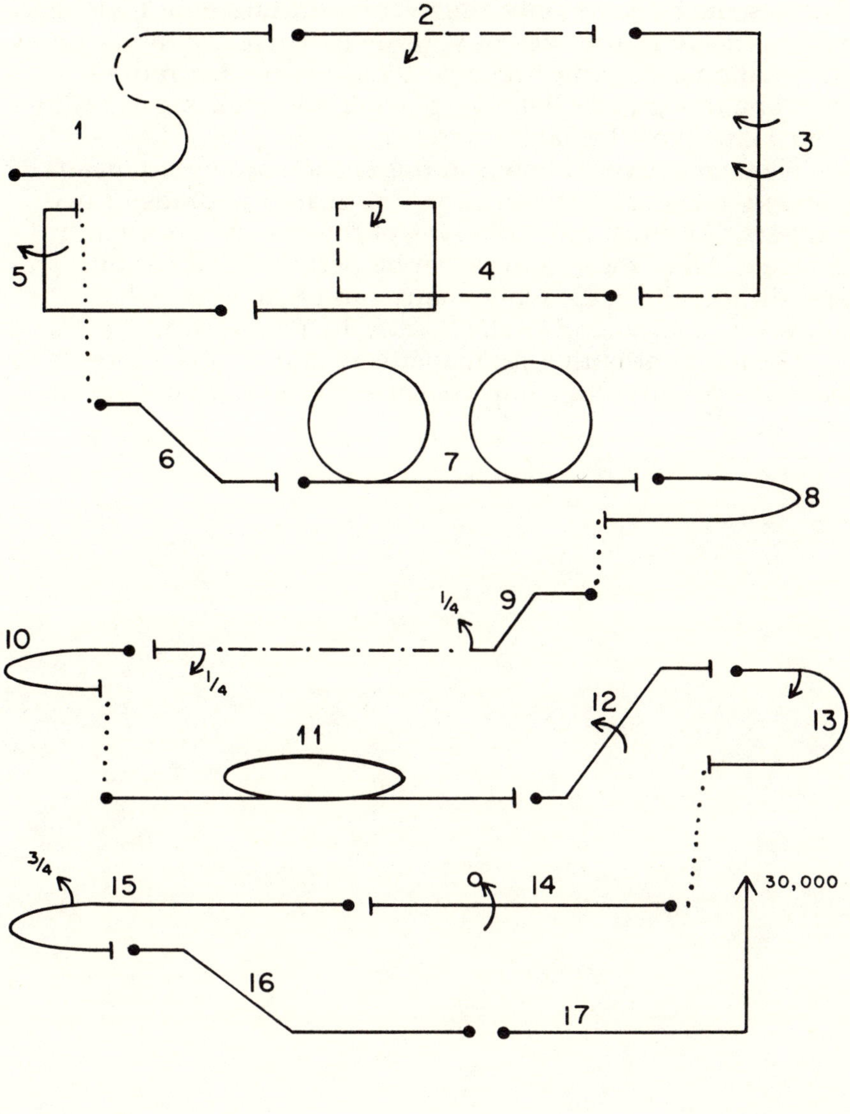

Sequence Flying

These sequences are only suggestions and training tools. Feel free to make up routines of your own. There are thousands of variations. Use the Aresti system to write down the maneuvers; this greatly helps in practice and makes a record of your maneuvers for later reference.

The maneuvers shown in this section are by no means the only ones you can perform. There are hundreds of maneuvers, but most are variations of the maneuvers already covered. All of these moves can be performed at various speeds and altitudes, while many can be flown in what is known as a *dirty configuration,* with the gear down.

Experiment with wild maneuvers and combinations. You have the privilege of complete safety. Why not take advantage of it?

CHAPTER 5
Two-Plane Aerobatics and Maneuvers

A number of simulations allow two-plane operation. Some use a split screen so both pilots use the same computer screen divided in half. Others offer a modem-operation option that lets each player use a separate computer. These games are often advertised as providing for two-player, head-to-head dogfighting. What many players don't realize is that this option also lets them fly together and perform many two-plane aerobatic maneuvers.

Learning to fly aerobatic maneuvers is tough, but learning to coordinate these maneuvers and fly them with another plane introduces a whole new level of complexity. Even if both pilots know the maneuver well, it can become fatally fouled up if they don't communicate well and are not used to flying with another plane close by. This type of flying does not come naturally. It must be practiced.

The best way to become acquainted with two-plane formation flying is to do it. First, you should master finding each other and flying side-by-side. This is called the *line abreast formation* or *combat spread.* Once you have the hang of forming up abreast and maintaining the same speed (this isn't easy), assign one pilot as *lead* and play follow the leader. Changing altitude together is simple but it won't take too long to figure out that staying together while turning is a problem. It will be well worth your time to practice and learn the following methods for turning in formation, or section reversal.

- The in-place turn
- The cross turn
- The split turn
- The tactical turn

These turns should be called by the leader. For instance, if the leader is going to make a tactical turn to the left, he or she should inform the other pilot, "Tac turn left."

Figure 5-1. Four Basic Two-Plane Maneuvers

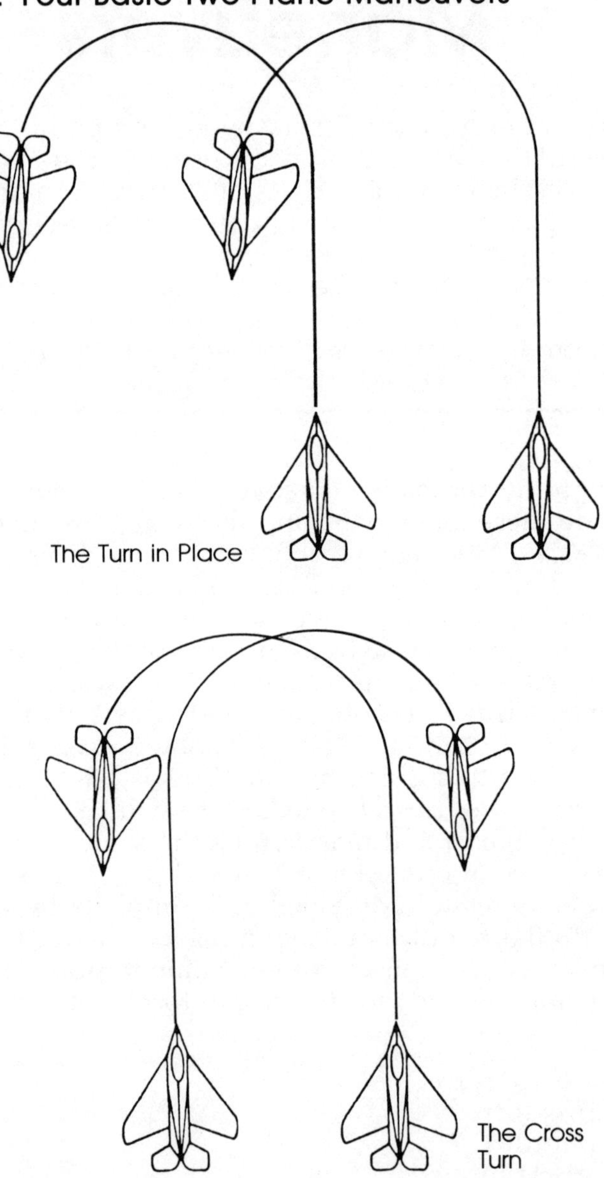

Figure 5-1, *continued*

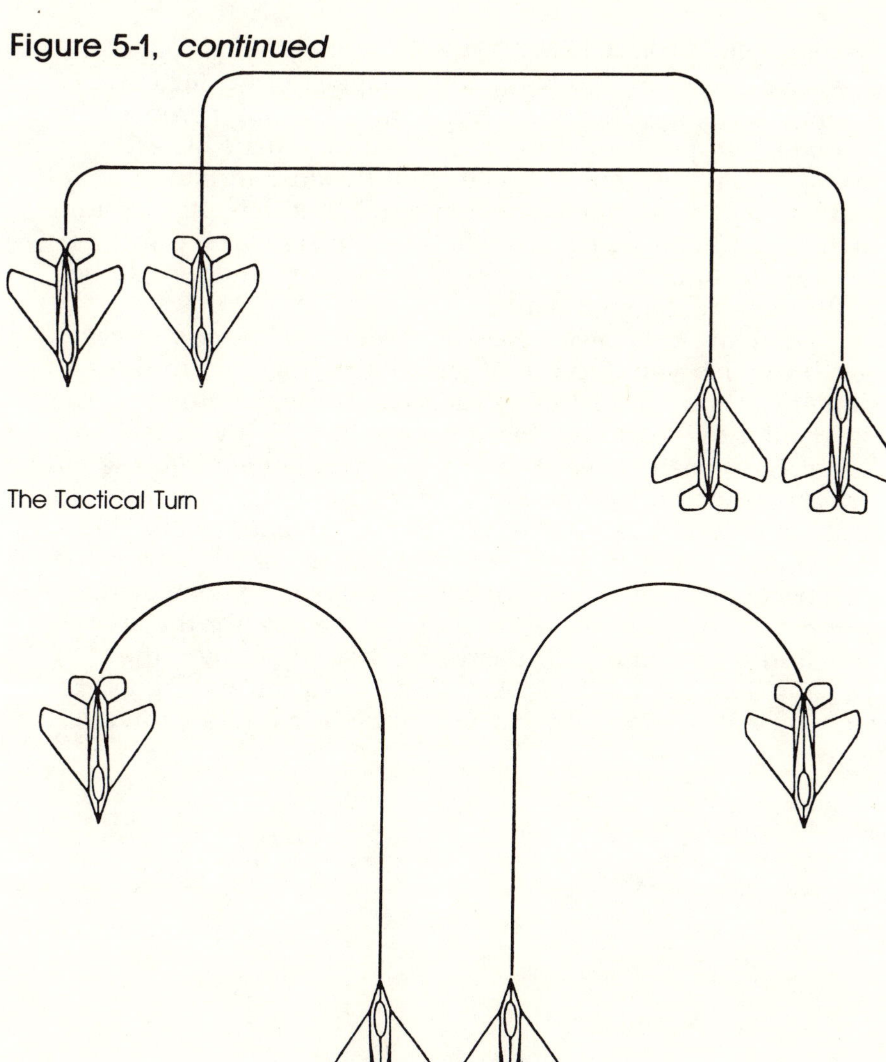

The Tactical Turn

The Split Turn

Two-Plane Aerobatic Maneuvers

If you've ever seen the Navy Blue Angels or the Air Force Thunderbirds in action, you know how exciting aerobatics can be when performed by more than one plane. These teams feature a group of planes that fly close formation maneuvers and two opposing solo planes that make the exciting high-speed head-on passes. The maneuvers covered in this section are similar to the types of maneuvers that would be performed by the solo pilots.

There are two basic types of maneuvers: *tandem maneuvers*, in which you and the other pilot fly together in the same direction, and *opposing maneuvers* in which you and the other pilot fly head-on toward each other. Flying in tandem has already been covered, but before you launch into the maneuvers it would be a good idea for you and your partner to practice several head-on passes. Some simulations allow midair collisions and some don't, so allow for enough space between your planes if collisions are possible. Remember that you're putting on a show, so pick a landmark for the show area and try to time your passes so that you're over the viewing area as you pass. A runway or tower make good reviewing areas, especially if your simulation offers a tower view.

Figure 5-2. The U.S. Navy Blue Angels in Tight Formation

The Head-On Pass

This is a basic maneuver you should master. There aren't usually minimum speeds for these maneuvers, but slower speeds allow more time for reaction.

- ⊙ Two planes start at opposite ends of the reviewing area. (1)
- ⊙ Fly toward each other and pass while over the viewing area. (2)
- ⊙ Pull up and turn back towards the reviewing area for the next maneuver. (3)

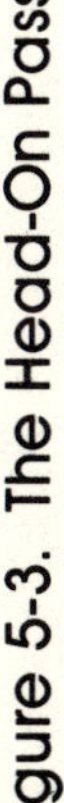

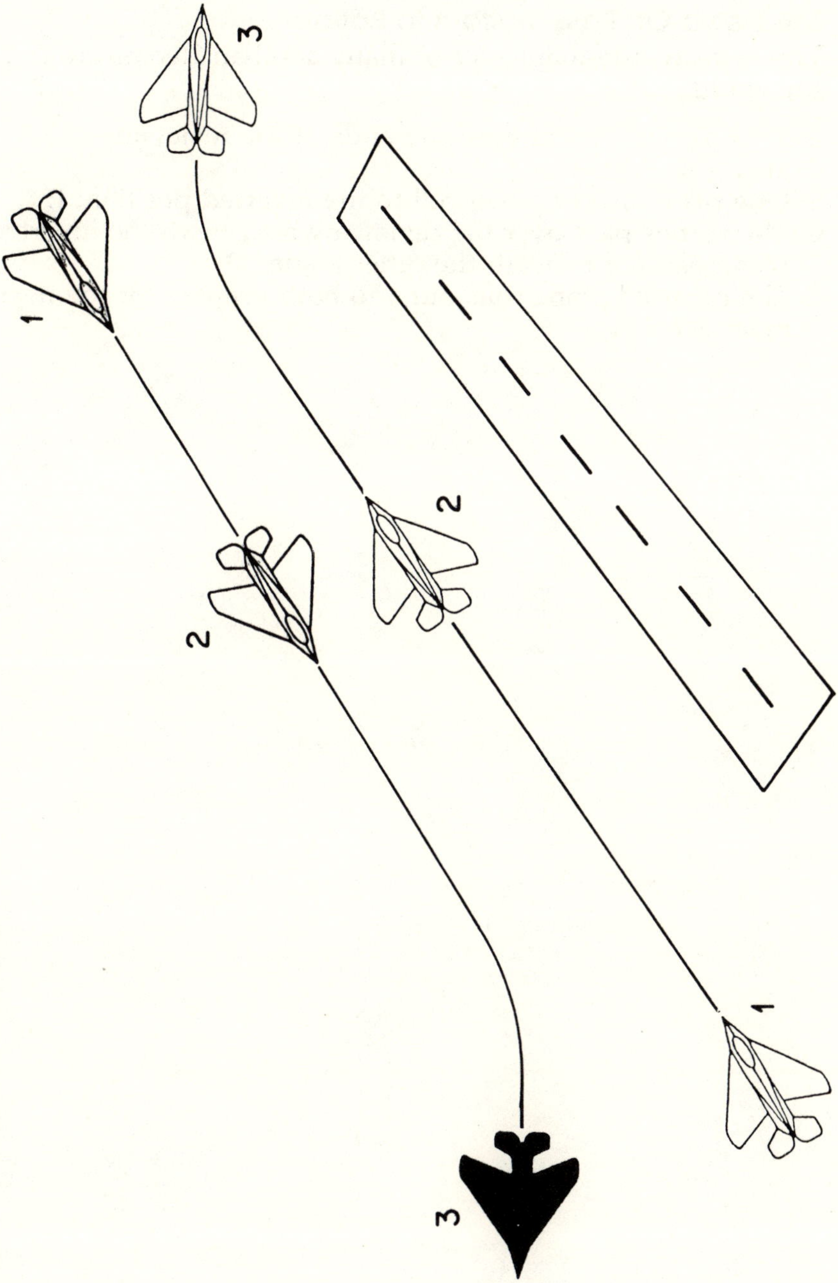

Figure 5-3. The Head-On Pass

The Head-On Pass: Bottom to Bottom

This is a favorite maneuver of many aerobatic teams around the world.

- ⊙ Two planes start at opposite ends of the reviewing area. (1)
- ⊙ One plane makes a half roll to the inverted position. (2)
- ⊙ The planes pass over the reviewing area, with the inverted plane passing beneath the other plane. (3)
- ⊙ The inverted plane rolls out and both prepare for the next maneuver. (4)

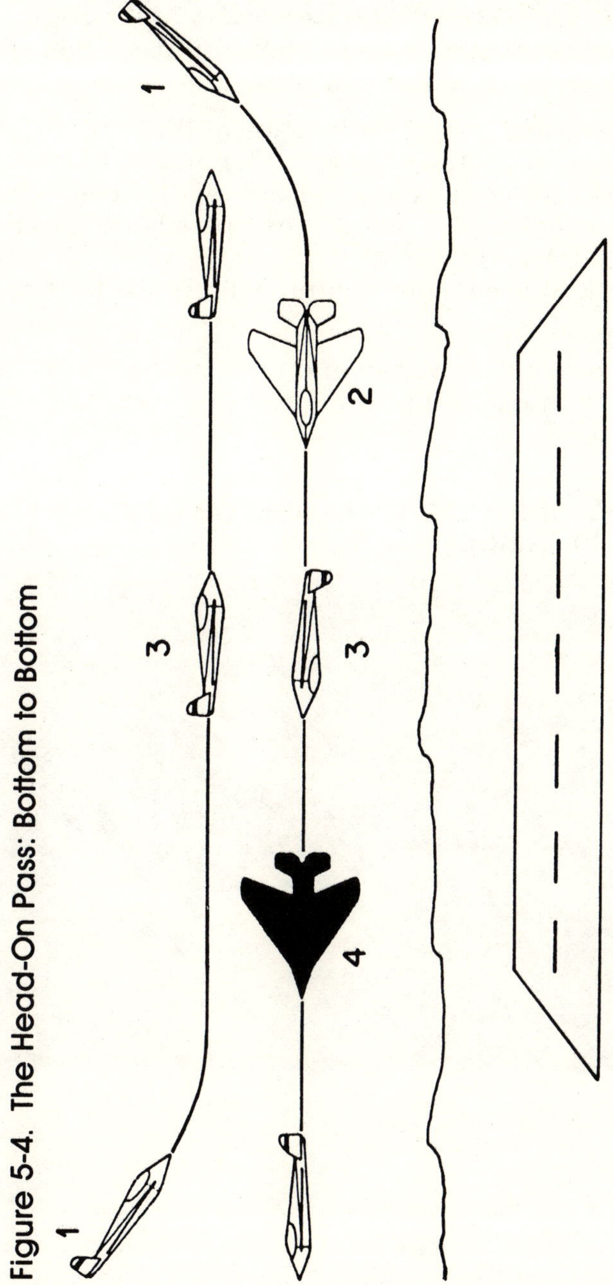

Figure 5-4. The Head-On Pass: Bottom to Bottom

The Head-On Pass: Bottom to Bottom, Knife Edge

This is another variation on the head-on pass. You should decide on which side you will pass.

- ⊙ Two planes start at opposite ends of the area. (1)
- ⊙ The planes head toward each other and just before they pass, each performs a quarter roll to a knife-edge position (on their right or left sides). They pass bottom to bottom on knife edge. (2)
- ⊙ Roll back to level flight and prepare for the next maneuver. (3)

There is another variation on this maneuver where, after the pass, both planes finish the rest of the aileron roll before pulling up.

Figure 5-5. The Head-On Knife-Edge Pass, Canopy to Canopy.

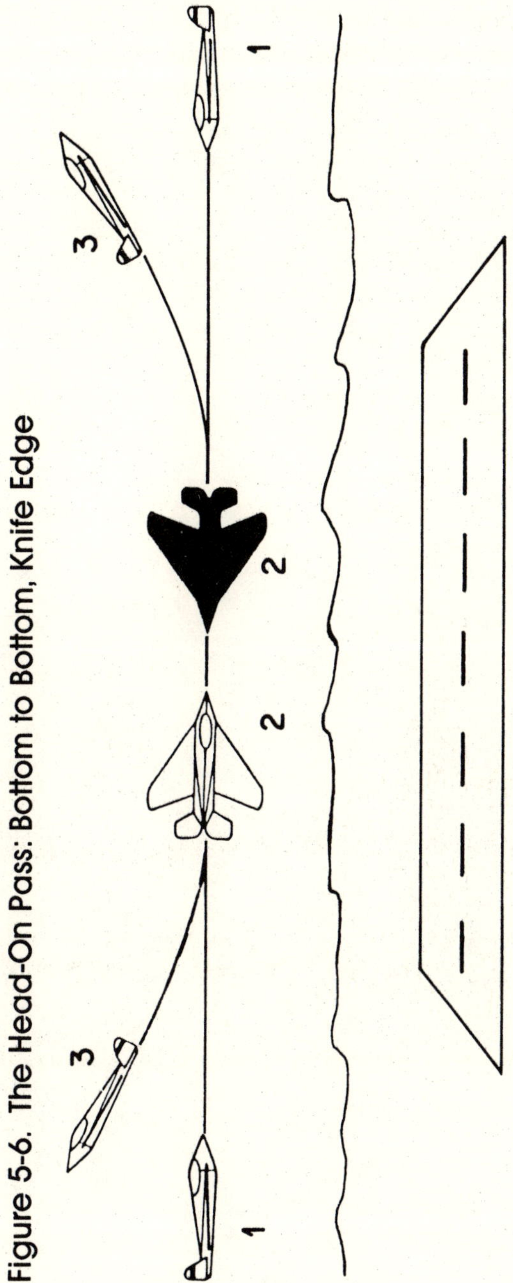

Figure 5-6. The Head-On Pass: Bottom to Bottom, Knife Edge

Head-On Loop

This is a maneuver that requires practice. If both planes don't fly loops of exactly the same size, they won't finish at the same time.

- ⊙ Two planes start at opposite ends of the reviewing area. (1)
- ⊙ As you pass, both planes pull up into loops. (2)
- ⊙ Both planes should finish the loop at the same point and altitude. (3)

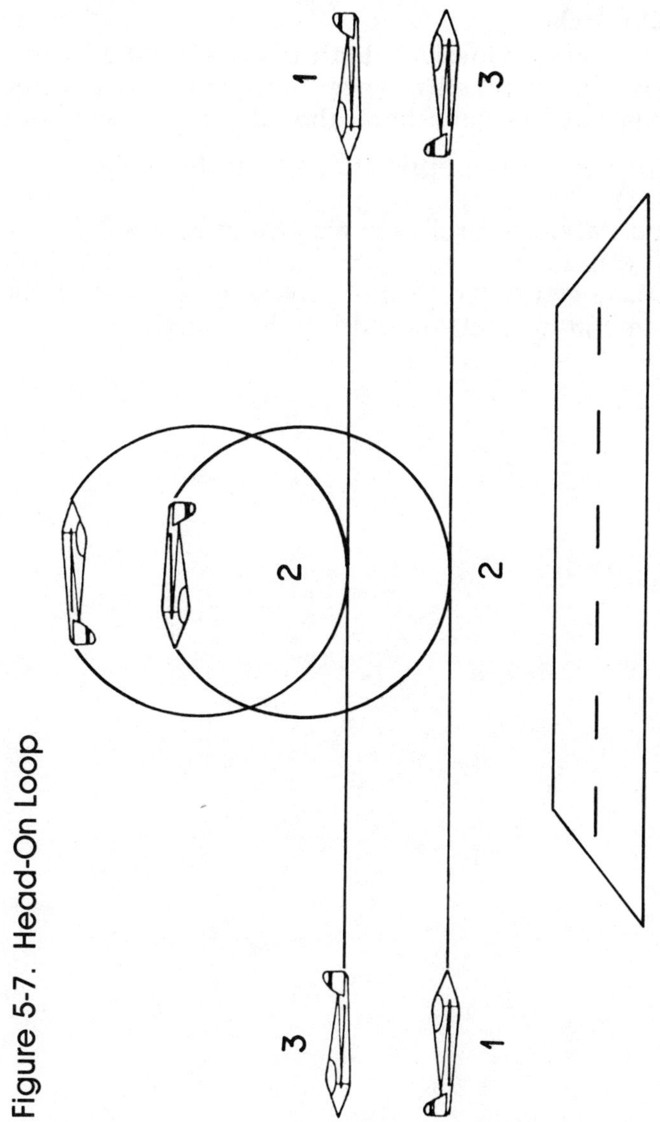

Figure 5-7. Head-On Loop

Head-On 360s

In this precision maneuver, both pilots should be very careful to have the same entry speed and bank angle or they won't finish at the same time, thus destroying the effect.

- ⊙ Two planes start at opposite ends of the reviewing area. (1)
- ⊙ As you pass, both roll into maximum-rate 360-degree turns. (2)
- ⊙ If all has been timed properly, you both should finish at the same point and roll out into level flight. (3)

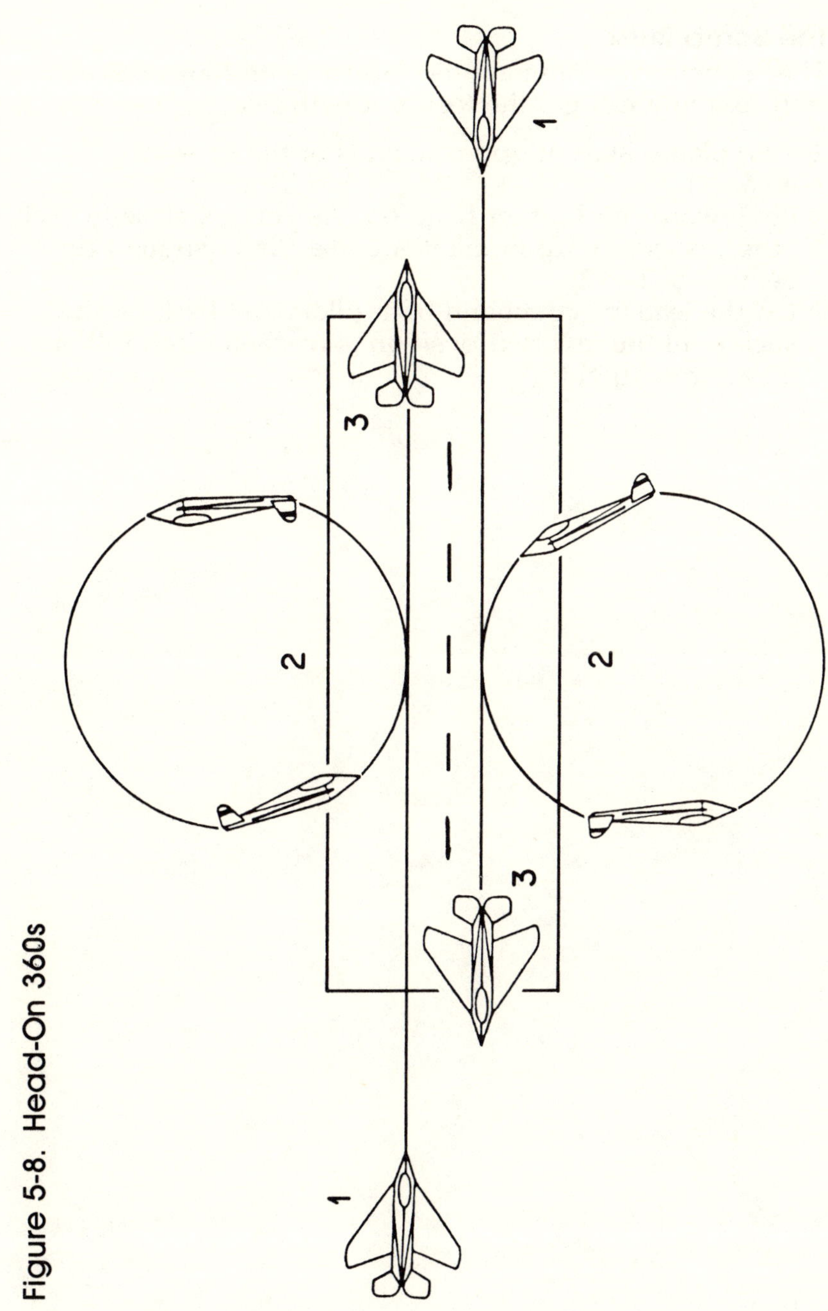

Figure 5-8. Head-On 360s

The Bomb Burst

This maneuver is normally performed with four or five aircraft, but you can get the feel of it with two.

- ⊙ Two planes start at opposite ends of the reviewing area. (1)
- ⊙ Fly towards each other head-on. As you get close to each other, both pull up into a 90-degree climb (straight up), belly to belly. (2)
- ⊙ On the leader's command both pilots pull back on the stick until the inverted position is reached, then roll out into normal flight. (3)

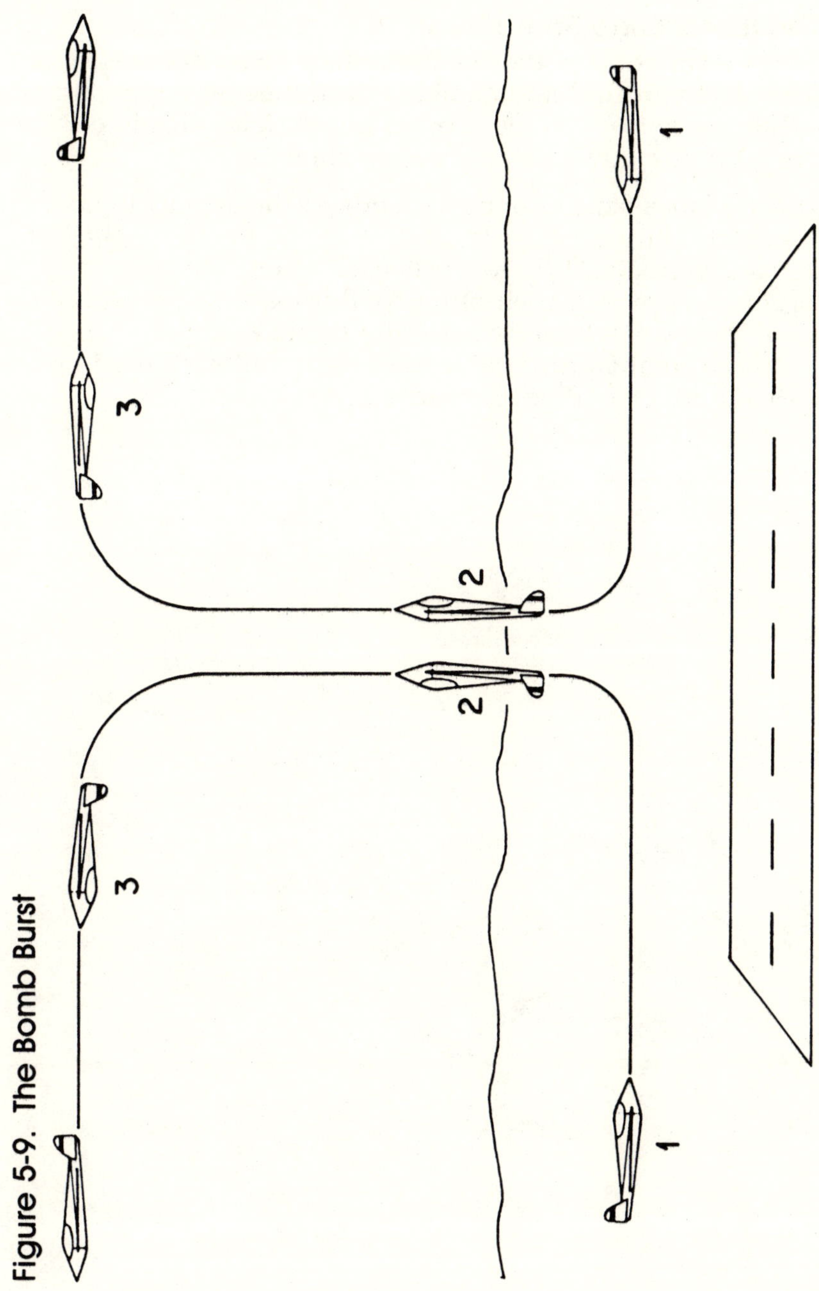

Figure 5-9. The Bomb Burst

The Descending Bomb Burst

This is a variation of the bomb-burst. Instead of meeting close to the ground and climbing, you meet at a high altitude and finish as close to the ground as possible. This is performed flying down as opposed to climbing.

- ⊙ Two planes start at opposite ends of the area at 10,000 feet. (1)
- ⊙ Roll inverted and fly towards each other. (2)
- ⊙ As you approach each other, pull back on the stick until you're in a 90-degree dive, belly to belly. (3)
- ⊙ On the command of the leader, both pull back hard on the stick until level flight is reached. (4)

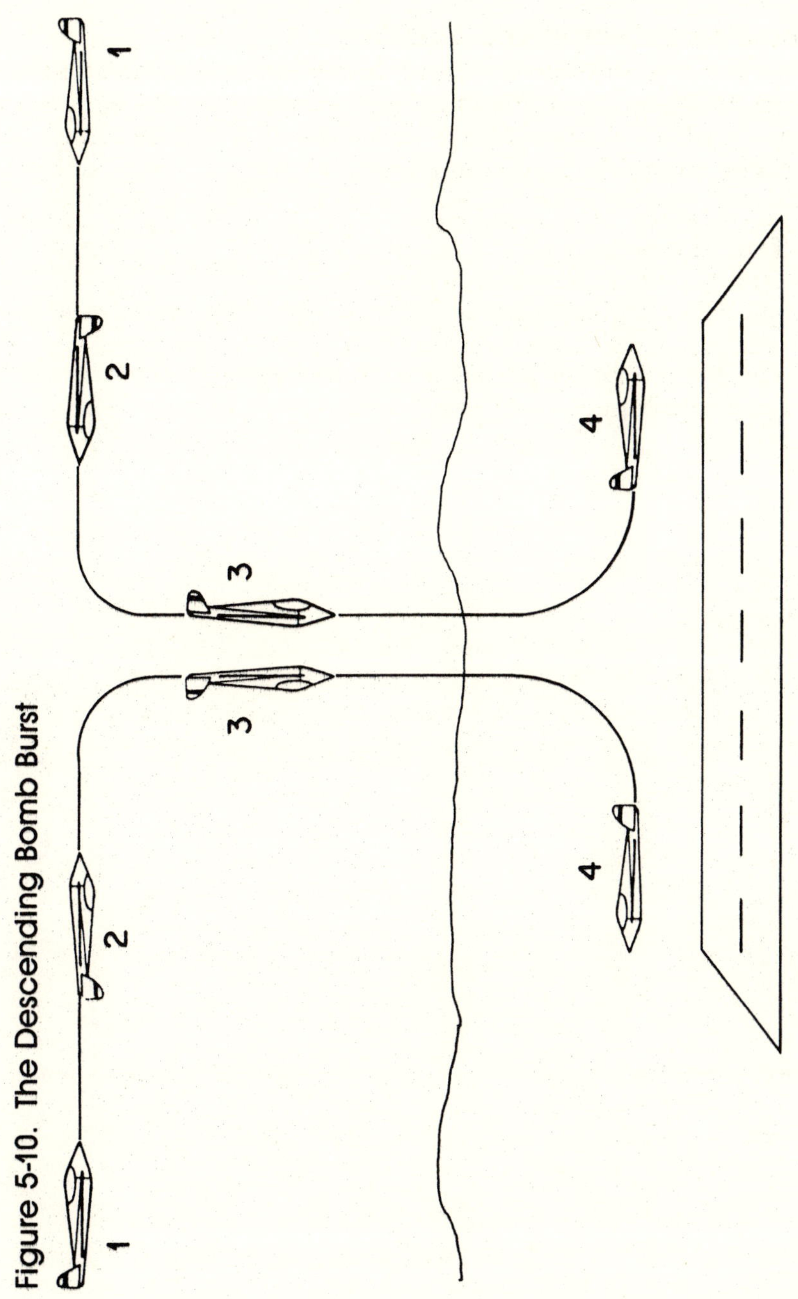

Figure 5-10. The Descending Bomb Burst

High-Speed Versus Low-Speed Pass

This is a great maneuver for showing off the possible speed range of the aircraft. Pilots in demonstrations are usually limited to submach speeds, but you have no such restriction. Depending upon the aircraft, you can show the difference between 180 knots or so, and mach 2.

- Two planes start at the same end of the area but one plane is much closer to the reviewing area.
- The closest plane flies as slowly as possible without stalling.
- The second plane passes just above the first plane flying at maximum speed with afterburners lit.

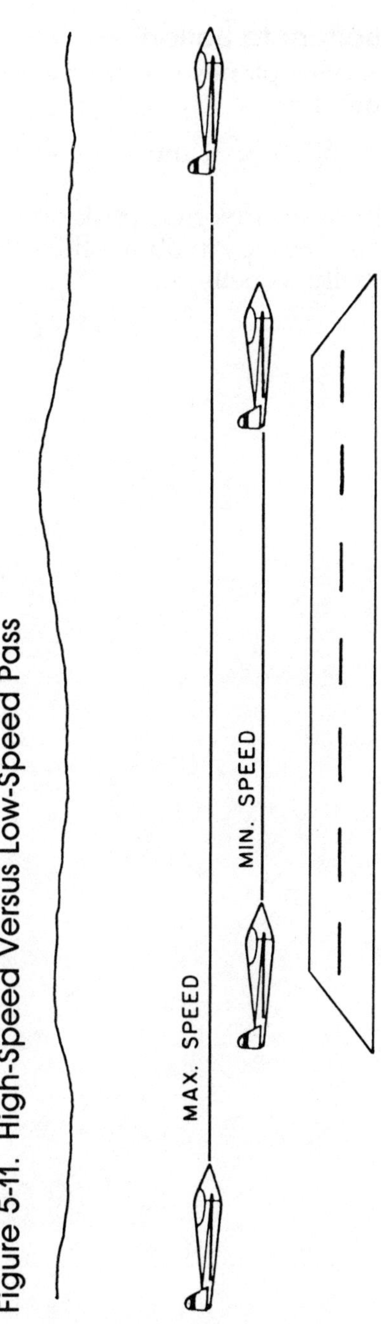

Figure 5-11. High-Speed Versus Low-Speed Pass

Tandem Pass: Bottom to Bottom

In a tandem pass both planes start at the same end of the field and close together.

- Both planes fly together at the same speed, from one end of the area. (1)
- The leader rolls to an inverted position. (2)
- The second plane moves in above the other plane so they're flying belly to belly. (3)

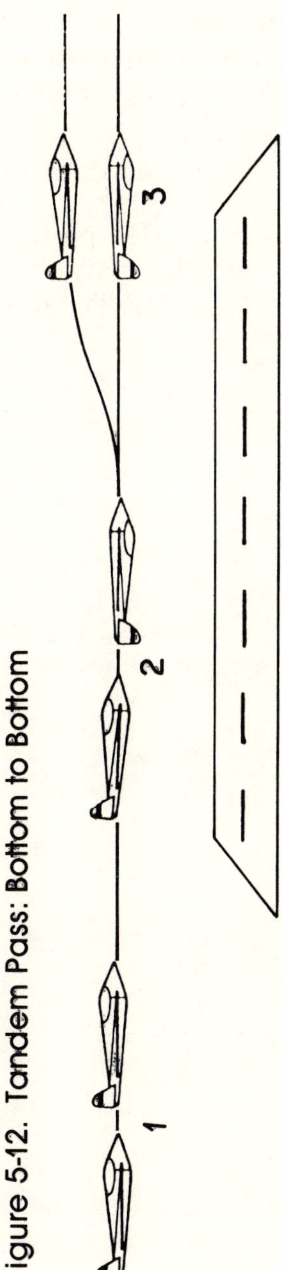

Figure 5-12. Tandem Pass: Bottom to Bottom

Tandem Pass: Top to Top

This maneuver is similar to the bottom-to-bottom pass, but it may be easier because the pilots should be able to see each other.

- ⊙ Both planes fly together from the same end of the reviewing area. (1)
- ⊙ The leader rolls to an inverted position. (2)
- ⊙ The second plane moves in beneath the leader and maintains the canopy-to-canopy position. (3)

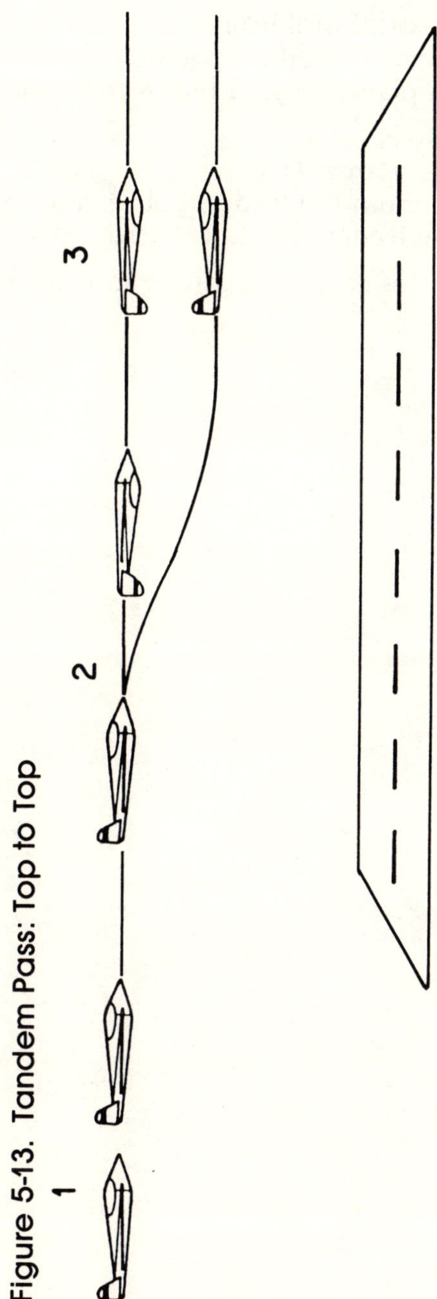

Figure 5-13. Tandem Pass: Top to Top

Tandem Pass: Four-Point Roll

Both planes fly close together, one behind the other in this maneuver. Both planes should roll in the same direction.

⊙ Both planes fly close together, nose-to-tail, from one end of the reviewing area. (1)
⊙ On command from the leader, both planes perform a four-point roll. The leader should call each quarter roll. (2)

Try to stay close together throughout the maneuver.

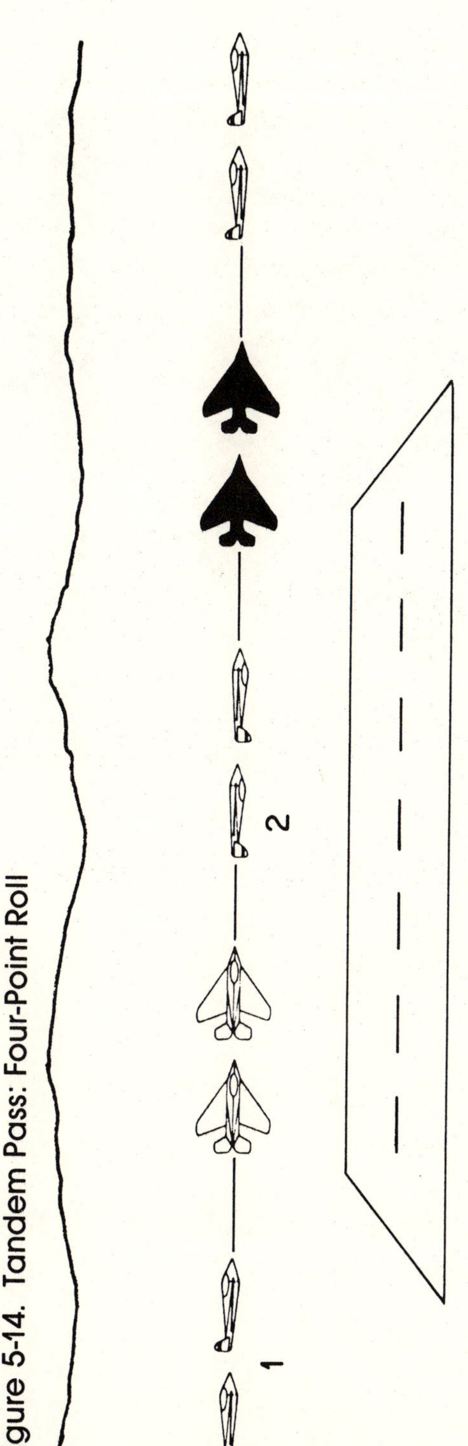

Figure 5-14. Tandem Pass: Four-Point Roll

Figure 5-15. Tandem Pass, Dirty (Gear-Down) Configuration

Side-By-Side Loops

To best perform this maneuver the wingman needs to keep sight of the leader. If your simulation offers a side view option, the wingman can fly the whole maneuver from this view. If this option is not available, the maneuver can be flown in the wing formation with the wingman slightly behind and to the side of the leader. In this manner the wingman can maintain sight of the leader during the whole maneuver.

- ⊙ Start side by side from one end of the area. (1)
- ⊙ When the middle of the reviewing area is reached, the leader should start a loop and the wingman should follow. (2)
- ⊙ Both planes should stay together and finish their loops at the same time. (3)

Figure 5-16. Side-By-Side Loops

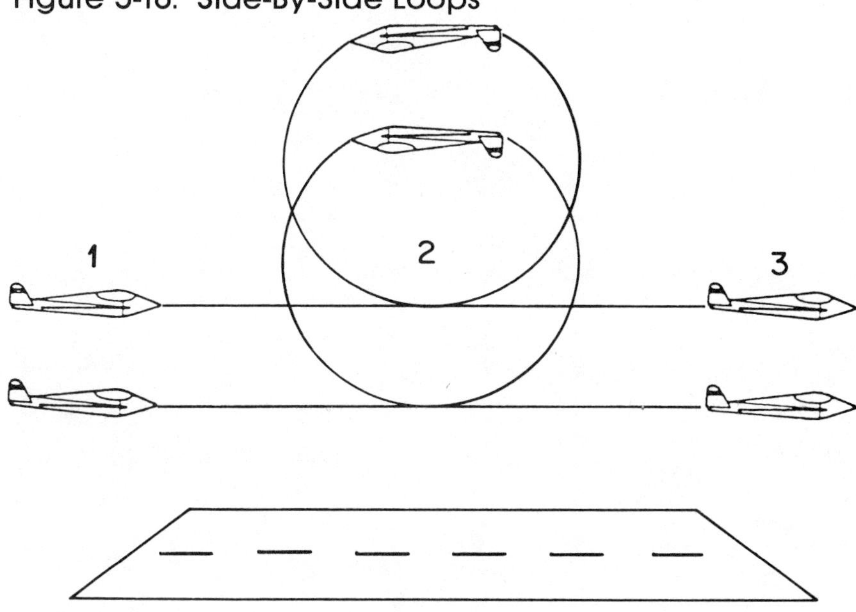

The Crossing Break

Take care during this maneuver to avoid colliding with your partner.

- ⊙ Both planes start together at the same speed, side by side with the leader slightly ahead. (1)
- ⊙ On the leaders command, both planes make a hard turn toward each other. The leader should pass in front. (2)

This maneuver can also be performed after a complete aileron roll for a more spectacular appearance.

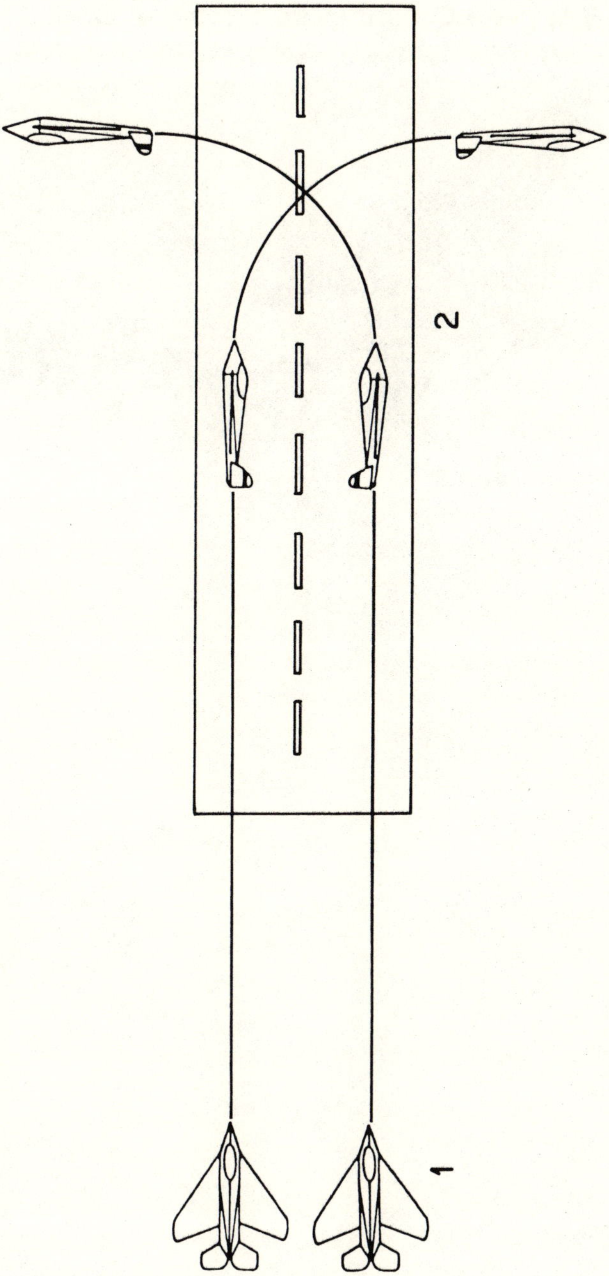

Figure 5-17. The Crossing Break

Figure 5-18. The Crossing Break, Phase One

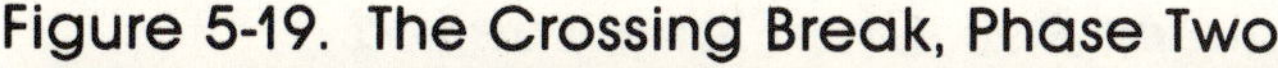

Figure 5-19. The Crossing Break, Phase Two

The Tandem Aileron Roll: One Plane Inverted

There are a number of variations on this maneuver. Learn this move and I'm sure you'll be able to think of many other similar maneuvers.

- ⊙ Fly nose to tail at the same speed from one end of the area. (1)
- ⊙ The trailing plane now rolls inverted. (2)
- ⊙ Both planes now perform an aileron roll, ending up as before with one plane inverted. (3)

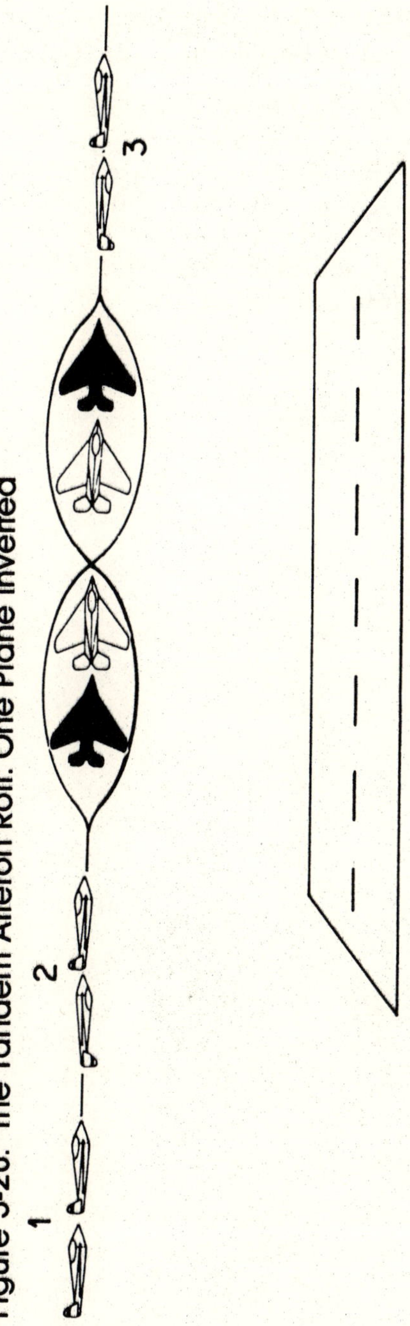

Figure 5-20. The Tandem Aileron Roll: One Plane Inverted

Summary

This is by no means intended to be a complete list of two-plane maneuvers. Once you learn these moves, you can start to modify them to make them more difficult. Add aileron rolls to your loops, or start your crossing break from the inverted position. The combinations are endless.

When you become expert at flying with a partner, put several of these maneuvers together into a routine. When you are able to fly two-plane maneuvers smoothly, you have truly mastered simulator flying.

Part 2

More Things to Do with *F-15 Strike Eagle*

CHAPTER 6
Red Flag Training for *F-15 Strike Eagle*

Just outside of Las Vegas Nevada, inside Nellis Air Force Base, is a small country called *Redland* which declares war on the U.S. Air Force five times a year and has yet to win. This is Operation Red Flag where pilots come to learn to fight their aircraft under wartime conditions and pressures. Although run by the Air Force, Red Flag participants can be from any of the U.S. armed forces as well as a number of foreign allies.

Red Flag was founded in 1975 as a result of the poor showing of Air Force pilots during the war in Southeast Asia. There, it became obvious that the training received by fighter pilots didn't go far enough. Studies showed that a pilot's skill (and his chances of survival) greatly increased during his first ten missions. After those ten, if he lived through them, he was as good as he was going to get. The problem was that too many pilots never made it to ten missions. Red Flag was founded to give pilots combat-like experience without having them risk their lives in real combat situations.

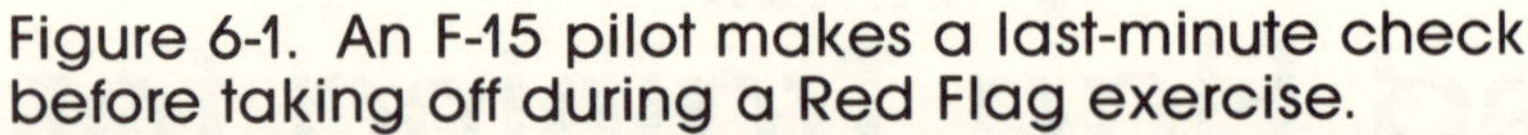

Figure 6-1. An F-15 pilot makes a last-minute check before taking off during a Red Flag exercise.

Each pilot at the school flies ten combat missions under extreme pressure, similar to that experienced in combat. During these flights he is encouraged to try new tactics and make mistakes; this is a learning experience, not an evaluation.

The air-to-air combat duels are flown against the pilots of the 65th Aggressor Squadron who simulate Soviet equipment and tactics. They fly F-5 aircraft to simulate the small, agile MiG. They also fly like Soviet pilots, which often gets them "killed" (or, in fighter pilot lingo, a *mort*). It takes a great deal of training and self restraint to be a good aggressor pilot. They often have to do things that aren't conventional in order to more closely resemble their Soviet counterparts. *Be Humble* is a motto they quickly come to appreciate.

F-15 Strike Eagle

Unlike the Navy's Top Gun school, Red Flag features air-to-ground as well as air-to-air missions depending on the role of the aircraft. Since you'll be flying the new *F-15E Strike Eagle,* you must excel in combat against ground forces as well as aircraft. A simulated Red Flag training program has been put together to help you learn to use all of the aircraft capabilities in a variety of situations. Training is divided into three parts, a section on air-to-air fighting one on one, a section on air-to-air fighting one on many, and a section on air-to-ground tactics.

Each mission, or *hop* as they're called, includes a premission briefing covering the mission setup, weapons to be carried, and flight goals. The mission goals will depend on the difficulty level you select. For normal difficulty, select the Pilot level. For extra difficulty, try the Ace level.

Start each mission according to the instructions given, even if they ask you to do some odd things, like letting yourself get shot. To complete a hop successfully you must fly safely back to the base. Bailing out does not count.

Air-to-Air, Part 1: Single Bogey

Red Flag Hop Number 1

Purpose: Hop number 1 is your qualifying flight for Red Flag. A certain level of proficiency is required to proceed with the rest of the missions. Proper completion this mission assures that you have reached this level.

Setup: Select *F-15 Strike Eagle* mission number 1—Libya.

Weapons To Be Used: All air-to-air weapons can be used.

Air-to-Ground Weapons: Drop all bombs at start of mission for maximum air-to-air performance.

Mission Description: Your mission is to fly southwest into enemy territory and engage any aircraft you encounter. Remember to allow yourself enough fuel to return safely to the base.

Mission Goals: Your goal for this mission is to shoot down ten enemy aircraft (Pilot mode), or seven enemy aircraft (Ace mode).

Red Flag Hop Number 2

Purpose: One of the problems encountered by pilots over southeast Asia was a reliance on missiles during air-to-air encounters. Once a fight became too close for missiles, they didn't have adequate skill in gun attacks to counter the maneuverable MiGs. This hop will force you to become acquainted with gun-only attacks.

Setup: Select mission Number 1—Libya.

Weapons To Be Used: For this mission only the gun may be used; the missiles should be carried but not used.

Air-to-Ground Weapons: All bombs should be dropped as soon as possible to achieve maximum maneuverability.

Mission Description: Fly southwest into enemy controlled territory, engage all enemy aircraft encountered, and return safely to the base.

Mission Goals: Your goal for this hop is to shoot down five enemy aircraft using only your gun in Pilot mode, or two aircraft in Ace mode.

Red Flag Hop Number 3

Purpose: The use of flares for countermeasures against enemy air-to-air missiles is no substitute for good evasive flying. This mission will help you gain confidence in evading enemy missiles without the use of flares. You will only be allowed to use short-range missiles so the enemy can get close enough to shoot back.

Setup: Select mission number 1—Libya.

Weapons To Be Used: Only short-range missiles and guns can be used during this hop. Flares may not be used.

Air-to-Ground Weapons: All bombs should be dropped as soon as possible.

Mission Description: Fly southeast into enemy controlled territory, engage all enemy aircraft, and return safely to the base.

Mission Goals: Your goal for this mission is to shoot down five enemy aircraft in Pilot mode, or three aircraft in Ace mode, and safely return to the base without the use of flares as a countermeasure against enemy missiles.

Red Flag Hop Number 4

Purpose: Your F-15 flies and handles beautifully under normal circumstances, but it can be a real bear to handle once you sustain damage. By completing this mission you will gain confidence in your ability to complete a mission even with an injured bird.

Setup: Select mission number 1—Libya.

Weapons To Be Used: Only short-range missiles and guns may be used during this mission.

Air-to-Ground Weapons: Drop all bombs as soon as possible.

Mission Description: Upon mission start-up, make a hard turn and allow the enemy aircraft to fire and hit you with one missile. Once you've been hit, you may engage and destroy the enemy aircraft that injured you and all other enemy aircraft you encounter. You must return safely to the base to successfully complete the mission.

Mission Goals: Your goal for this mission is to shoot down five enemy aircraft in Pilot mode, or three aircraft in Ace mode, after receiving damage.

Air to Air, Part 2: Multiple Bogeys

Fighting when outnumbered is never easy, but it doesn't have to be a disaster either. The main thing to remember is to stay on the offensive and cut the odds as quickly as possible. Also, it's important to avoid long engagements with any one particular aircraft because doing so might allow another bogey to swing around behind you for a good rear-quarter shot. Keep your energy level high and make a series of slashing attacks rather than long, turning engagements.

Red Flag Hop Number 5

Purpose:	This mission is designed to help you to determine your level of expertise in the area of multiple bogey fighting. You'll have unlimited use of all weapon systems and defensive countermeasures.
Setup:	Select mission number 5—Hanoi.
Weapons To Be Used:	All weapons can be used in this mission. A good tactic is to fire medium-range missiles as soon as contact is made in hopes of cutting the odds before the engagement actually starts.
Air-to-Ground Weapons:	Drop all bombs as soon as possible.
Mission Description:	Your mission is to fly west into enemy-controlled territory and engage all aircraft encountered. Return safely to the base.
Mission Goals:	Your goal for this mission is to shoot down eight enemy aircraft in Pilot mode, or five enemy aircraft in Ace mode.

Red Flag Hop Number 6

Purpose:	Now we're going to make things a bit tougher. You'll fly basically the same mission as before, but this time without the advantage of medium-range missiles. You'll be forced to really "mix it up" with the bogeys.
Setup:	Select mission number 5—Hanoi.
Weapons To Be Used:	Only guns and short-range missile may be used on this mission.

Air-to-Ground Weapons: Drop all bombs as soon as possible.

Mission Description: Your mission is to fly west into enemy-controlled territory and engage all fighters encountered, using only your guns and short-range missiles. Return safely to the base.

Mission Goals: Your goal for this mission is to shoot down six enemy aircraft in Pilot mode, or four enemy aircraft in Ace mode.

Red Flag Hop Number 7

Purpose: You can't always count on having a load of missiles, so you need to learn to use your gun in a multiple-bogey situation. Don't chase one plane long enough to let another get behind you. High speed frontal attacks work well. Disengage if you need to and then start the attack again. Use your afterburners to create some space between you and the enemy, if needed.

Setup: Select mission number 5—Hanoi.

Weapons To Be Used: You are to use only guns during this mission, but you can use all available countermeasures.

Air-to-Ground Weapons: All bombs should be dropped as soon as possible.

Mission Description: Fly west into enemy-controlled territory, engage all aircraft encountered, and return safely to the base.

Mission Goals: Your goals for this mission are to shoot down five enemy aircraft in Pilot mode, or three enemy aircraft in Ace mode, and return safely to the base.

Red Flag Hop Number 8

Purpose: Often your mission will include an air-to-ground component. To accomplish this part

	of the mission you may have to fight your way into the target, carrying a load of bombs. The performance of your aircraft is affected by the extra weight of the bombs, so it's critical that you learn to maneuver in combat with a heavy load.
Setup:	Select mission number 5—Hanoi.
Weapons To Be Used:	On this mission you may use short-range missiles and guns only.
Air-to-Ground Weapons:	Carry all bombs for the entire mission.
Mission Description:	You are to fly west into enemy-controlled territory, engage aircraft, and return safely to the base.
Mission Goals:	Your goals for this mission are to shoot down six enemy aircraft in Pilot mode, or four enemy aircraft in Ace mode, and return to base.

Part 3: Air-to-Ground

The F-15E was specifically designed to be an all-weather attack fighter capable of assorted air-to-ground missions. This being the case, two Red Flag training missions are included to help you hone your bombing skills in the combat environment.

Red Flag Hop Number 9

Purpose:	Low-level high-speed bombing is a skill that should be mastered by all attack pilots. This mission gives you the opportunity to practice this skill.
Setup:	Select mission number 3—Haiphong.
Weapons To Be Used:	No air-to-air weapons will be needed; countermeasures may not be used against missiles.
Air-to-Ground Weapons:	Use bombs to attack ground targets.

Mission Description: You're to fly west into enemy-controlled territory and attack both primary targets flying below 1500 feet on afterburners. Cut afterburners just before you reach the target, pop up to 3000 feet, and dive on the target. Attempt to hit each target with more than one bomb. After you hit the first target, turn on the afterburners and attack the second target. Maintain an altitude below 1500 feet except when popping up to dive on the target. Return safely to the base.

Mission Goals: Hit both primary targets. This should give you a score of 3000. Try to hit at least one with two bombs, giving you a score of 4500 in Pilot mode.

Red Flag Hop Number 10

Purpose: The other type of bomb run that every attack pilot should know is the dive-bomb attack. In this mission you will enter the area above SAM SA-7 range at 35,000 feet and turn on afterburners. The heat-seeking missiles can't reach you and the radar-guided missiles can't catch you. Dive on the target, using the speed brake to keep from exceeding the Vmax for the aircraft.

Setup: Select mission number 6—Iraq.

Weapons To Be Used: Air-to-air weapons will not be needed. Countermeasures can be used when leaving the area and diving on the target; they should not be needed when entering the area.

Air-to-Ground Weapons: Use bombs on ground targets.

Mission Description: Fly east into enemy-controlled territory and attack the first SAM (Surface to Air Missile) site. Fly at an altitude of 35,000 feet on afterburner. Set the NAV (Navigation) cursor to direct you to the target. When the blinking aircraft cursor enters the NAV box, cut the afterburners and dive straight down on the target. It will be necessary to extend the air brake to avoid exceeding the Vmax. Use the medium-range radar to locate the target on the way down, make any adjustments necessary, and pull up at 5000 feet. Then drop your bomb on the target. Learning this technique may take several tries, so don't be surprised if your first few attempts are failures. The most common error is starting your dive too late and passing over the target.

Mission Goals: Successfully dive-bomb one of the SAM sites around the closest airfield and return safely to the base.

CHAPTER 7
More Missions for *F-15 Strike Eagle*

One of the best things about *F-15 Strike Eagle* is that each mission includes a tenable scenario that adds a little flavor to the action. You're flying an F-15 across the so-called "line of death" into the Gulf of Sidra. You will face Libyan pilots in their MiG-23s and SU-22s; this setup is much more exciting than flying against pilots and targets from an imaginary country labeled only as *the enemy.* It captures your interest right away. And if you aren't careful you just might learn something.

With that in mind I have written five more scenarios for *F-15 Strike Eagle*. Some are factual and some are (to use a favorite term of the intelligence community) entirely *notional.*

Each mission includes a preflight briefing, mission goals, and suggested flight plans. I hope you have as much fun flying them as I had writing them.

Good hunting.

Operation Swift Kick

Use F-15 mission number 1—Libya.

Preflight Mission Briefing: After a couple of years of relative calm, Colonel Qadaffi and his band are up to their old tricks again. After the U.S bombing raid in 1986, Qadaffi pleaded with the Kremlin for "humanitarian aid" to help clean up the significant mess F-111 and A-6 attack planes can make. Unfortunately, most of the money he received to help the sick and injured went to the purchase of a number of Chinese Silkworm missiles. The Silkworm is the Chinese equivalent of the much-feared Exocet, which was used with great effect to sink the HMS Sheffield during the Falklands War.

Qadaffi could hardly wait to try out his new weapons. Last month he started firing them at random components of the U.S. fleet operating in the Mediterranean. Fortunately, the crews were on guard and defeated the missiles by either shooting them down or fooling them with chaff. Some F-14 crews were even lucky enough to locate the MiG-23s used to launch the missiles and made short work of them.

Undaunted, Qadaffi continued the attacks. This morning, however, the colonel made a serious mistake. His pilots ventured out into the darkness again to seek out the U.S. fleet. A large ship was located where they had found the fleet several days before and two missiles were sent skimming across the wave-tops. The 864 sleeping passengers and crew members of the cruise ship *Dalmatia* had no warning. Both missiles struck amidships at the waterline. Thick, black smoke prevented many passengers and crewmembers from making it to the lifeboats. Forty-eight minutes after being hit, the ship capsized. Rescue efforts are continuing.

Such irresponsible actions can no longer be tolerated. A swift and serious response has been called for. This mission has been designed with two objectives:

- To eliminate Qadaffi's ability to launch air strikes by destroying all airfields.
- To demonstrate our superior power by accomplishing this mission with a single aircraft.

Resistance is expected to be low due to the rapid nature of our response.

Mission Goals: Your F/A-18 will be launched from the deck of the *USS Ranger*. With it, you will attack all of Libya's coastal airfields and return safely to the carrier.

Suggested Flight Plan:

Figure 7-1. Suggested Flight Plan for Operation Swift Kick

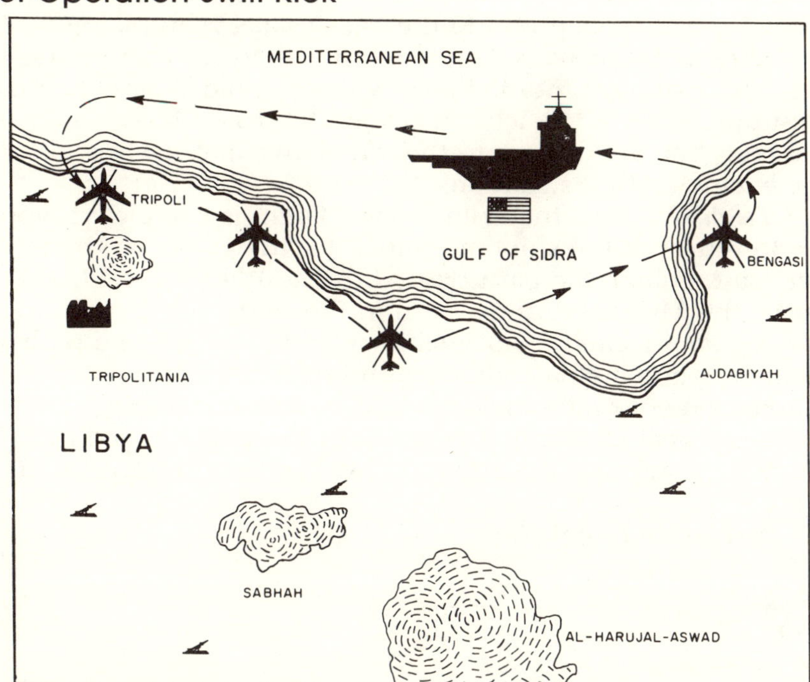

Operation Wild Weasel

Use F-15 mission number 3—Haiphong.

Preflight Mission Briefing: The month is April, 1967. This is a two part mission.

Flight 1: January of 1967 went very well for the U.S. Air Force over North Vietnam. A number of MiGs had been downed with few losses. Then the rainy season set in and few missions were flown. The enemy has taken advantage of the overcast sky to move large numbers of troops and quantities of war materiel. Antiaircraft systems have been improved around most major targets. The number of MiGs in the area has increased.

When the clouds cleared in April, and the bombing raids resumed, pilots started having problems. In dogfights during the first several days of April, the Air Force shot down nine MiGs, but lost seven aircraft in the process. Due to the increased losses, the Joint Chiefs of Staff have authorized an attack on the MiG bases and they are allowing the Air Force to attack the MiGs while still on the ground. An attack on Kep MiG Base, deep into Route Pack 6, has been planned.

Due to the increased antiaircraft activity around the target, the first flight into the area will be a Wild Weasel Flight to suppress the SAM activity around the base. This is your mission: You are to fly into the target area and destroy all the enemy SAM installations.

Flight 2: Once the Wild Weasel Flight is completed, you are to return to the ship and reload for the attack on the target itself. Follow the path created by the Wild Weasel flight, attack the MiG base, and return to the ship.

Mission Goals: Your goal for Flight 1 is to clear a path to the target, eliminate all surrounding SAM locations marked on your mission map, and return to the ship.

Your goal for Flight 2 is to follow the path created by Flight 2, attack and destroy the MiG base, and return to the ship.

Suggested Flight Plan:

Figure 7-2. Suggested Flight Plan for Operation Wild Weasel

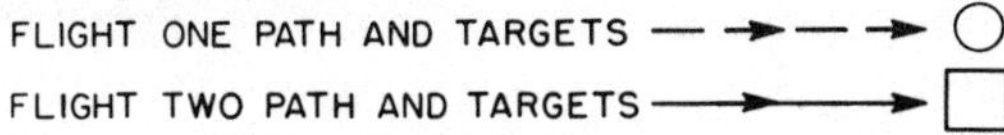

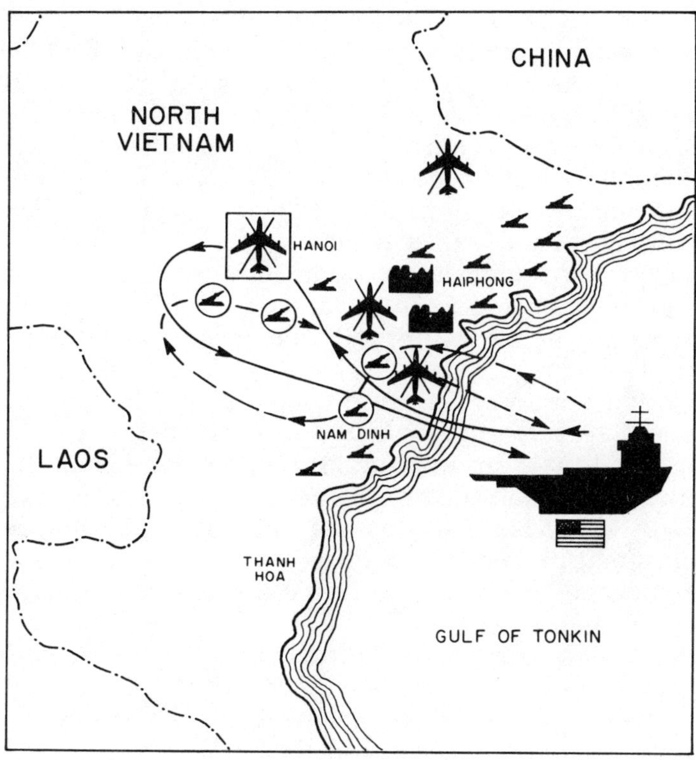

Operation Beach Blanket Bingo

Preflight Mission Briefing: As large as the U.S. Fleet's presence is in the Persian Gulf, it can't be everywhere at once. This morning, after several days of Iranian gunboat attacks on tankers passing through the Straits of Hormuz, the fleet was moved south to the mouth of the straits to protect shipping in the area. Now, with the fleet in the south, the Iranians have launched an assault across the gulf onto the beaches of Kuwait. Their apparent aim is to establish a beachhead in Kuwait and move inland into the populated areas before the U.S. fleet can react. If they can accomplish this they will force the Iraqis to split their army in order to protect their border with Kuwait.

There is also another problem. Through excellent intelligence or just dumb luck, the attack was timed perfectly so our air wing is at the end of a flying cycle. All of the aircraft are out of fuel and the tankers have landed. Nothing in the air is loaded with air-to-ground weapons anyway. We do have one F/A-18 Hornet on board being loaded and fueled.

Fly this F/A-18 up the gulf and attack the Iranian force while they're still on the beach. You will be carrying the new CBU-87/B cluster bombs. This is a three-in-one ordnance, combining antipersonnel, antiarmor, and incendiary effects in one package. A hit with a couple of these should stop the invasion in its tracks.

You'll need to conserve your fuel as this will be a long flight. A short burst of afterburner, however, can be effective against the slower heat-seeking SAMs. Tensions in the area are high, so expect to be fired upon from both sides of the Gulf; Iranian planes may also attack. Your best bet is to avoid the Iranian coast on the way up so you can't be intercepted. After your bomb run, you may drop all remaining bombs to lighten your load and reduce fuel consumption.

Mission Goal: Your goal for this mission is to drop two bombs on the assigned target in one pass and return to the carrier. If intercepted by Iranian aircraft you may defend yourself, but don't go looking for a fight on the way to the target: Time is critical.

Suggested Flight Plan:

Figure 7-3. Suggested Flight Plan for Operation Beach Blanket Bingo

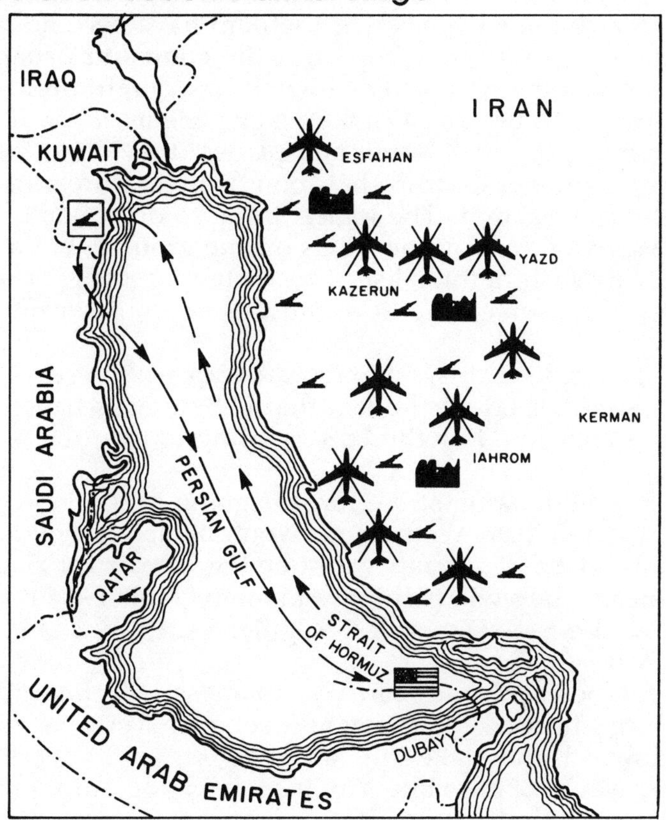

Operation Bolo

Preflight Mission Briefing: The date is January 2, 1967. The bombing raids over North Vietnam have become predictable. U.S. jets fly in from the same direction and usually at the same time every day. The enemy has caught on to the process and has begun to cause real problems. Even if they don't manage to shoot down any planes, they often force the incoming bombers to drop their bombs prematurely in order to take evasive action. The logical thing to do is bomb the MiG bases and destroy the MiGs on the ground, but airbases are still on Washington's list of restricted targets. There has to be a way to force the MiGs into the air so they can be fought.

To accomplish this, an elaborate trap will be set. The enemy has capitalized on the fact that U.S. attacks have become predictable. Now the U.S. is going to turn that against them.

A normal-looking strike force is going to the North today as usual. The North Vietnamese won't be able to tell from their radar that the aircraft won't be big, heavy F-105 bombers. Instead, they will be highly maneuverable F-4 Phantoms armed with air-to-air missiles. Regular F-105 call signs and tactics will be used.

If all goes well, the North Vietnamese will think they have F-105s headed their way and send up MiGs to shoot them down. If they take the bait, it's your job to shoot down as many MiGs as possible. You'll probably be outnumbered so make your shots count. Remember that you may have to fight your way out, so reserve a missile or two for that possibility.

Mission Goal: Your goal for this mission is very straightforward: You're to fly west into North Vietnam and shoot down MiGs until you're out of weapons or fuel. With any luck you should become an Ace (five kills) during this mission. There's a good possibility that you could become a double Ace if your flying and shooting skills are topnotch.

Suggested Flight Plan:

Figure 7-4. Suggested Flight Plan for Operation Bolo

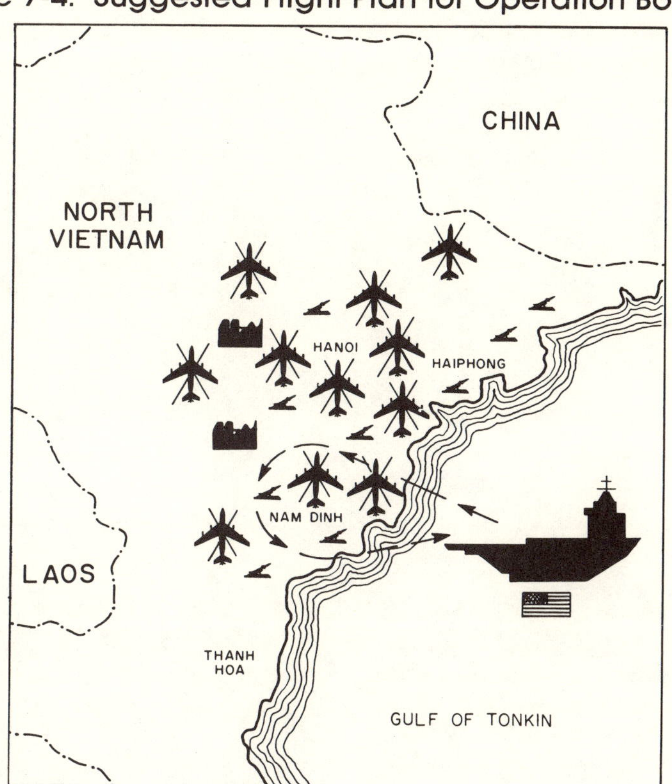

Part 3

Tips and Hints for Jet Fighter Games

I make a hobby out of buying, collecting, and playing new games, especially games involving jet fighters. As such, I am well aware of the frustration accompanying the first couple of game sessions with a new simulation. You have to learn a whole new set of controls and procedures and how the program reacts to your inputs, much the same as a pilot would do when getting checked out in a new aircraft. But the frustrating part usually comes when you go out to face the enemy. Each programmer creates intelligence for the enemy differently. Tactics that work in some games won't work in others and you're often left staring at the screen wondering what happened as the computer makes a smart remark about notifying next of kin.

In this section you'll find some insight into how the games are set up in terms of what works and what doesn't, and how what you do or don't do affects your score.

CHAPTER 8
Tips and Hints for Playing *Stealth Mission* by SubLOGIC

Stealth Mission is the latest release from the founder of the flight simulation movement, SubLOGIC. There are a number of mission scenarios and a wide variety of difficulty levels. Interesting control tower and spot plane views are offered and should be familiar to *Flight Simulator* and *Jet* fans. This is a good and complex game and as such, is somewhat confusing at times. I hope the following tips will allow you to enjoy the game more and increase your scores.

Aerobatics

The X-29 is an outstanding aerobatics aircraft that features forward-swept wings. It can perform outside loops (be careful of negative G forces) and turns and handles very well. Like *Jet,* this game lets you use the scenery disks that are also produced by SubLOGIC. With these disks you can perform your air show over hundreds of different airports all over the world. The folks at SubLOGIC tell me that with a new disk about to be released, you will even be able to defect and fly your super secret F-19 fighter to Moscow. You will have to locate it yourself, however: The airport is on the disk, but no electronic navigational aids are provided.

Note Maximum Weight Before Loading

Before you start to load the aircraft with ordnance, note the maximum weight the aircraft can carry. This may effect your selection of weapons.

Maximum Takeoff Weights are as follows:

Aircraft	Maximum Takeoff Weight	Can Carry as Ordnance
F-19	33070 lbs	11020 lbs
X-29	17800 lbs	4000 lbs
F-14	74349 lbs	34245 lbs

Although the game is named after the F-19 Stealth Fighter, due to its outstanding maximum speed, range, and payload capability, the F-14 will likely be your choice of aircraft for many missions

Air-to-Air Weapons Default

When you first start up, air-to-air weapons will be preselected. As a result, ground targets won't appear on the Raw Scope. To avoid flying into trouble, you should cycle the weapons control over to an air-to-ground weapon. If you do this, vectors to all ground targets will show up on the screen, letting you plan your attack.

Target Hits

Some targets may not be destroyed by a single bomb hit, depending on the ordnance. But this is not necessarily bad if the target is not a primary target. If you are attacking a missile frigate or SAM site you only need to put the radar out of action; a single hit will normally do that. You will also get credit for a hit on the screen.

The Wonderful Shift-G

Straight and level flight is something that any pilot should be able to accomplish without much thought, but many flight games make it very difficult to do this. The programmers of *Stealth Mission* have eliminated this hassle with the Shift-G command. If you hold down the Shift key and press the G key, your plane will automatically stabilize on a level flight. Why waste time giving the joystick bumps in one direction, then in the other for five minutes, trying to achieve level flight when you can accomplish the same thing with a single keypress. This command frees you from some of the tedious flying adjustments and lets you concentrate on the more complicated areas of the game. Learn to use it.

Plan Your Attack

As you approach enemy territory you can usually tell a great deal about the enemy deployment from the radar. Usually it's a bad idea to barrel straight into a target area. Plan your attack to minimize your exposure to SAM sites. This can be done by attacking across the enemy front rather than into it. If you head straight in you may come under fire from one target while you're attacking another, and if you turn to avoid that attack you may be detected by yet another SAM site. If you attack across the front, you can generally line up the first set of targets and take them out one at a time. Also, you'll always know which way to turn to get away from the threat area.

Figure 8-1. Attacking Across the Enemy's Front

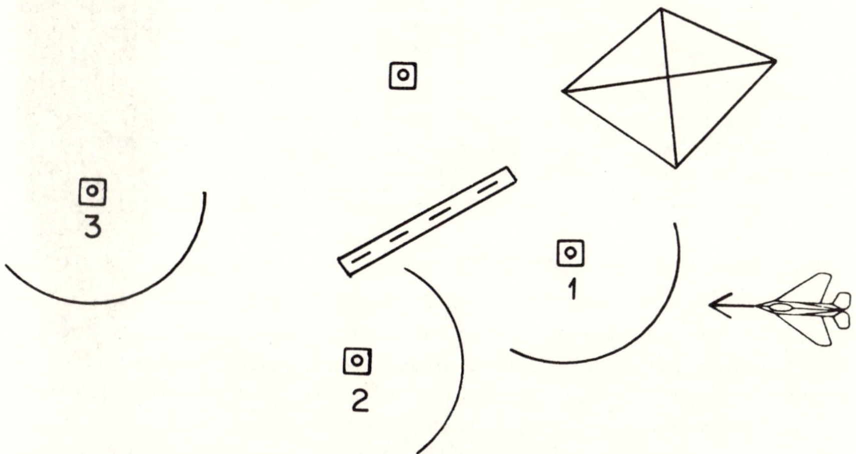

When attacking across the front, you can take out targets 1, 2, and 3 in order and always know you have a safe area to your left.

Figure 8-1. Attacking Across the Enemy's Front, ***continued***

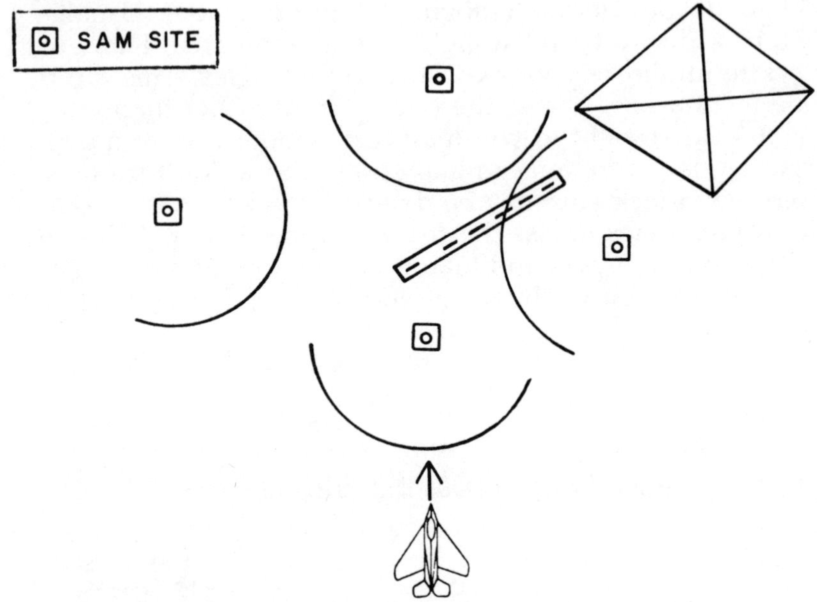

If you attack straight into the target area, you are heading into the teeth of enemy defenses. You may come under fire from several directions at once and have limited escape routes.

Break Away After Bombing

The main idea when attacking a ground target is to drop your bombs before you come into range of the target's defenses. It will take several seconds for the bomb or missile to reach its target. If you continue to fly straight toward the target after dropping, you will likely wind up with a SAM on your tail as a result. One good tactic is to drop your bomb and then make a quick 360-degree turn. By the time you come back around to the target you should know if it was a hit or a miss.

Hit or Miss?

Sometimes it seems as if it takes forever for a missile to reach its target. Before you assume that it's a miss, use the *missile's eye view* command. If you get an image, the missile has not reached the target. It still might be a hit.

Another Miss!

Speaking of missed bombs, you may wonder why so many low-altitude drops come up short. When you lock on a target, the effect of gravity will often cause bombs dropped in level flight to fall short. The manual says to aim a little beyond the target to compensate, but that's hard to judge. The best way is to drop your bombs in a shallow dive. If you are approaching the target at low altitude, pop up briefly to gain enough altitude to safely dive on the target.

Target Locked

The best way to insure a target hit is to get a good lock on the target before you drop. Your accuracy will vary greatly depending upon the magnification level you're using. 8× magnification level will give the best results when aiming for lock up; using 16× causes too large a jump on the screen when you move the joystick. A target may look perfectly lined up when on 1× magnification, but switching to 8× will often reveal that the aim is far from true.

Figure 8-2. At 1× Magnification the Target Appears to Be Lined Up

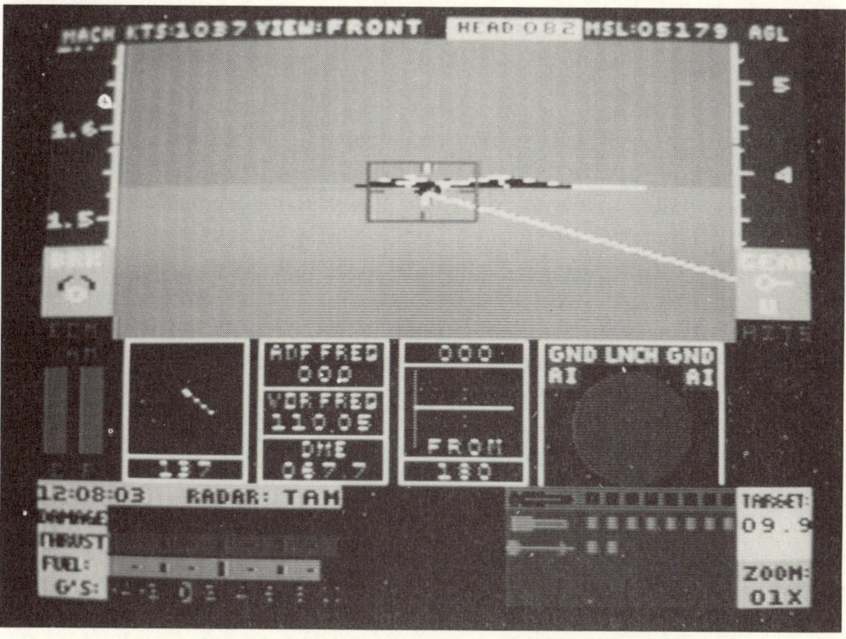

Figure 8-3. 8 × Magnification Reveals That Your Missile Will Hit Mountain, Not Bridge

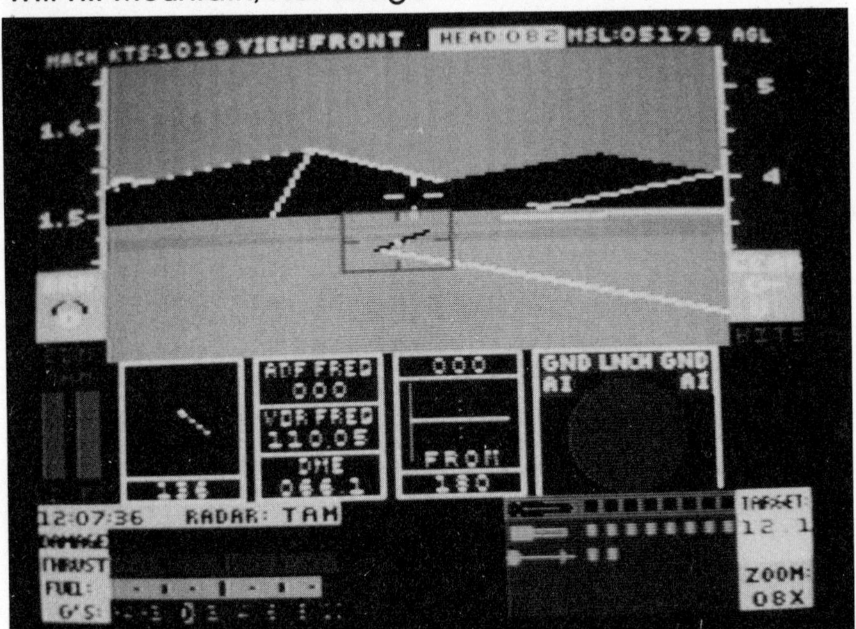

Locking Problems

The current version of the game has one small problem: If you move the joystick in any manner after you lock onto a target, your weapon will be launched. The folks at SubLOGIC say they are aware of the problem and it should be corrected in future versions. Until then you can avoid the problem by using the keyboard for course corrections after locking or waiting until you are ready to launch to lock on the target. By using the keyboard, you can lock onto several targets while you're a safe distance away and then launch as you get closer, thus attacking several targets at once.

Crosshair Color Change

In mountainous areas you will have a tough time aiming with the crosshair sight since both the sight and the mountains are black. One option is to use the O key to switch the scenery to wire-frame mode, but the designers went to a lot of trouble to provide solid graphics so it would be a shame to defeat them. The other option is to use the U key to change the color of the crosshairs. You will be offered a series of color options; choose the one that gives you the best contrast.

First Mission

Your first mission will be full of new things to learn, so start with something less dangerous, like the Mountain Conflict—Easy. This mission provides a number of features that will help you learn quickly to fight in the *Stealth Mission* environment.

First, there are a number of excellent terrain features that can be used to navigate, such as roads, rivers, and mountains. Learning to pay attention to the terrain and navigate using its features will become increasingly important as the difficulty level rises.

The second interesting feature of Mountain Conflict—Easy is the location of the first two primary targets. The two bridges are the first things encountered as you head into enemy territory, and they are lightly defended. This will give you the opportunity to try out your weapons systems and perfect your target aiming-and-locking skills.

The SAM sites will only fire on you if you approach within a certain area. This area increases as you increase in difficulty. To avoid being shot, you should fire your weapons from as far away as possible, then turn away from the target.

Flares and Chaff

Unlike their real-life counterparts, the aircraft in *Stealth Mission* aren't equipped with radar warning receivers or infrared tracking receivers. This equipment would tell you that you've been locked onto by enemy weapons and what type of missile is coming at you. In this game you never know if a SAM is radar-guided or heat-seeking. Consequently, you don't know which countermeasures to employ.

Fortunately, the game designers have supplied you with a large number of both types of countermeasures, so when fired on by a SAM, it's best to use both types just to be sure. Don't deploy them just once. Flares and chaff only last a couple of seconds, while a SAM may chase you for a minute or more. Keep dropping flares and chaff, and keep jinking the aircraft until the missile threat is gone. Use the radar jammer also.

Think Twice Before Dropping That Last Bomb

When you have air-to-ground weapon selected, the raw scope will provide information regarding the location of ground targets. Once you drop your last air-to-ground weapon, you'll no longer be able to get that information. If a number of SAM sites are between you and the base, you'll want to know where they are in order to avoid them on your way out. If you have only one target and one bomb left, by all means drop the bomb. If a large number of targets remain, and you are going to make a second trip anyway, save the last bomb for the last target on the way out.

Be Careful Around the Mountains

In computer games, depth perception is severely limited. You lack good visual clues to tell you how high an object is. When flying close to mountains, keep this in mind. It's very difficult to tell when you're higher than the peak of a mountain. Give yourself plenty of extra clearance or go around the mountains.

Magnified Views

Using the magnified view option when targeting and navigating can be very helpful, but remember that you're seeing a magnified image with a limited field of view. If you forget your view is magnified, you may think you're beyond a mountain when, in fact, you are not. You're at an increased risk of running into something. When not targeting or trying to find your way around, use the 1× or 2× view.

Take Care at Low Speeds

When landing or flying at low speeds, remember that any maneuver you make will cause an additional loss of airspeed. This could cause a fatal stall during landing.

Remote Control Flying

Using the control tower view, you can fly the aircraft of your choice as if it were a remote control plane. Flying this way is difficult because you will lack an artificial horizon indicator.

When the plane is far away you'll have trouble seeing its attitude. Remember to use the Shift-G function to level out the wings. Use to ADF to fly the plane back toward the tower.

Low Fuel Hint

If you're low on fuel or have a damaged tank, you'll want to maximize your fuel economy when flying to the base or tanker. One way to reduce your fuel consumption is to reduce your weight, but the game does not offer a jettison-weapons feature. To get rid of your load it will be necessary to locate and lock onto a target, then drop all your weapons on it. Air-to-air missiles will probably have to be carried along unless you are attacked.

Fighting Speed

When dogfighting you want to be able to turn as quickly as possible. The speed at which most fighters turn the best is around 500 knots. Try to maintain this speed during dogfights. You may need to throttle back at first, adding throttle as you bleed off speed maneuvering.

Using VOR as a Landing Aid

The manual gives a fair description of VOR and what it is. The manual doesn't explain the value of the VOR, or how to use it, however. You don't really need it to find the base since the ADF will do that for you. You can use it, however, to help line your aircraft up for landing. To do this you will need to know the direction in which the runway is laid out. The best way to find this is to note your heading when you take off straight down the runway. You can then use the VOR to fly straight back down that radial when you want to land. By doing so you will be perfectly lined up on the runway.

The easiest way to explain the procedure is by example. Check your manual and make sure you know where the VOR omni-bearing indicator is located. Then go to the Mountain Conflict mission and get ready to take off.

You should be lined up on a heading of 000.

- Your OBI needle should be in the center of the instrument and the OFF indicator should be on.

- ⊙ Take off and maintain your heading of 000. Fly up to 2000 feet. The OBI should now indicate FROM 180.
- ⊙ Continue to fly north until your DME shows 50.0. You are now 50 miles north of the base. (You can use the afterburners to speed things up.)
- ⊙ Now make a 180-degree turn to the left. Make your heading 180 degrees. You are now heading south towards the base. Set the OBI to TO 000.
- ⊙ Your OBI needle should now be slightly to the left of center. Cut your power to half and fly towards the needle until the needle is in the center of the area. Make small maneuvers and wait for the needle to move. The OBI reacts slowly so give it a chance to start moving.
- ⊙ The needle will probably move toward the center and keep on going to the right. You flew over the 000 radial, but you aren't on it yet.
- ⊙ Keep adjusting your heading until the needle is in the middle of the area. When the ADF is pointing straight up and you are on a heading of 180 degrees, you're flying straight down the 000 radial in line with the runway.
- ⊙ To get all these things to line up you'll need to anticipate your turns. If your turn to the left has the needle moving to the center, turn back to the right just before it gets there to avoid an overshoot.
- ⊙ Don't worry if you have trouble the first time. Do it until you get the feel of it. It will become second nature shortly.
- ⊙ Fly on over the runway at 2000 feet. You may have to make some last-minute adjustments to get directly over the strip. At this point it's easier to use the magnified view option to line up in the runway than it is to chase the needle.

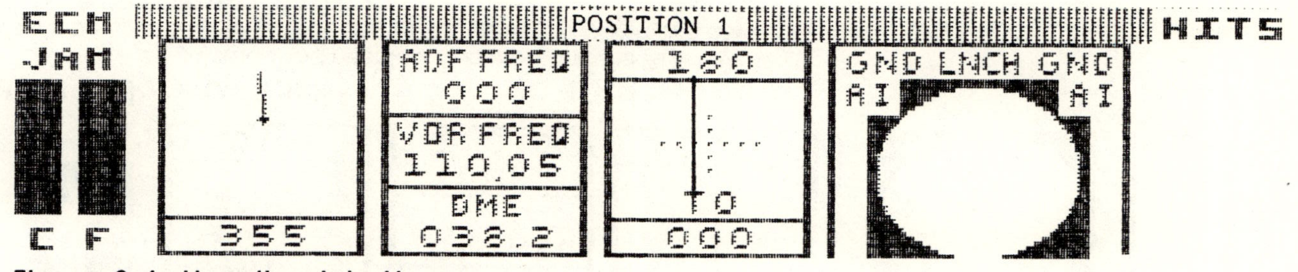

Figure 8-4. Heading into Home
You have turned toward the base, and are heading in the right direction. The OBI needle shows that a line extending from the end of the runway is to your left. Bank to the left. When the needle gets close to the center, turn back to the right until your heading is 180.

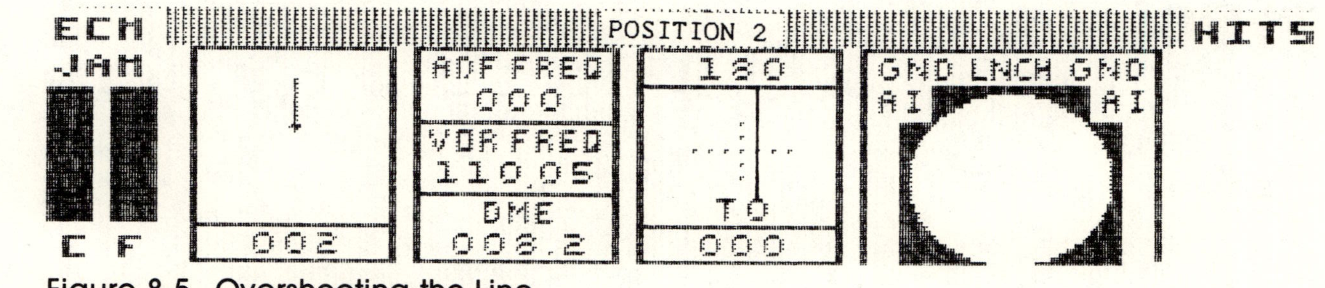

Figure 8-5. Overshooting the Line
New pilots often overshoot the line and the needle moves past center to the right of center, as it is in this figure. You are close to the right heading, but minor adjustments must still be made.

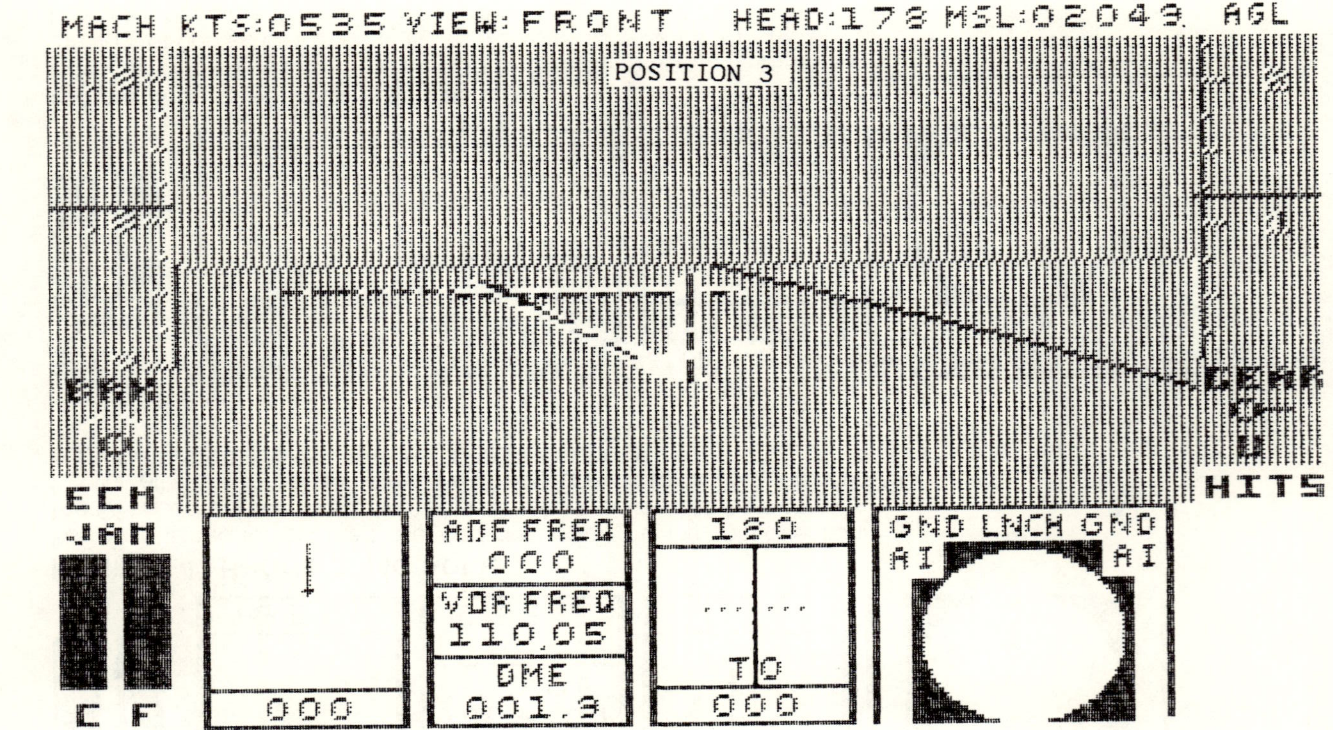

Figure 8-6. Back on Track
Now you're on the right track. The needle is centered and the ADF is pointing straight up. As the runway comes into view, you can make any minor course corrections as necessary.

Touch and Go

The runways on which you must land seem awfully narrow; putting the aircraft down straight on one takes skill and practice. Military pilots practice by doing touch-and-gos. To do a touch-and-go, come in for a normal landing and when your wheels touch the ground, apply full power and take off again. This is often repeated a number of times by flying a race track pattern where you touch down, take off, and go around. The following procedure will help you master the touch-and-go; use the Mountain Conflict again.

- Line up for a landing using the VOR set for TO 000 degrees as we did before.
- Gradually reduce power, speed, and altitude until you're about 50 feet up and at 300 knots airspeed.
- Use the magnified view options to line up the runway. Once over the runway, lower your gear, reduce power to zero, and touch down.
- Immediately after touchdown, apply full power and take off again.
- Make 180 degree turn to the left. Note your bank angle because you'll want to bank again at the same angle later. The best way to do this is to either bank at 90 degrees (totally sideways, making it difficult to level out at the right heading), or bank so the horizon goes from corner to corner on your screen.
- Your heading should now be 000 degrees. Continue on this heading until you are 15 miles from the base. You can reduce this distance as your skill improves.
- When you're 15 miles from the base, make another 180 degree turn (the same angle as the first) and make your heading 180.
- Now start your descent and line up on the runway as you did before.
- Land, and repeat the procedure.

Once you can perform a series of touch-and-go landings without crashing or landing too far from the runway, landings should never be a problem again.

Figure 8-7. Touch and Go Circuit

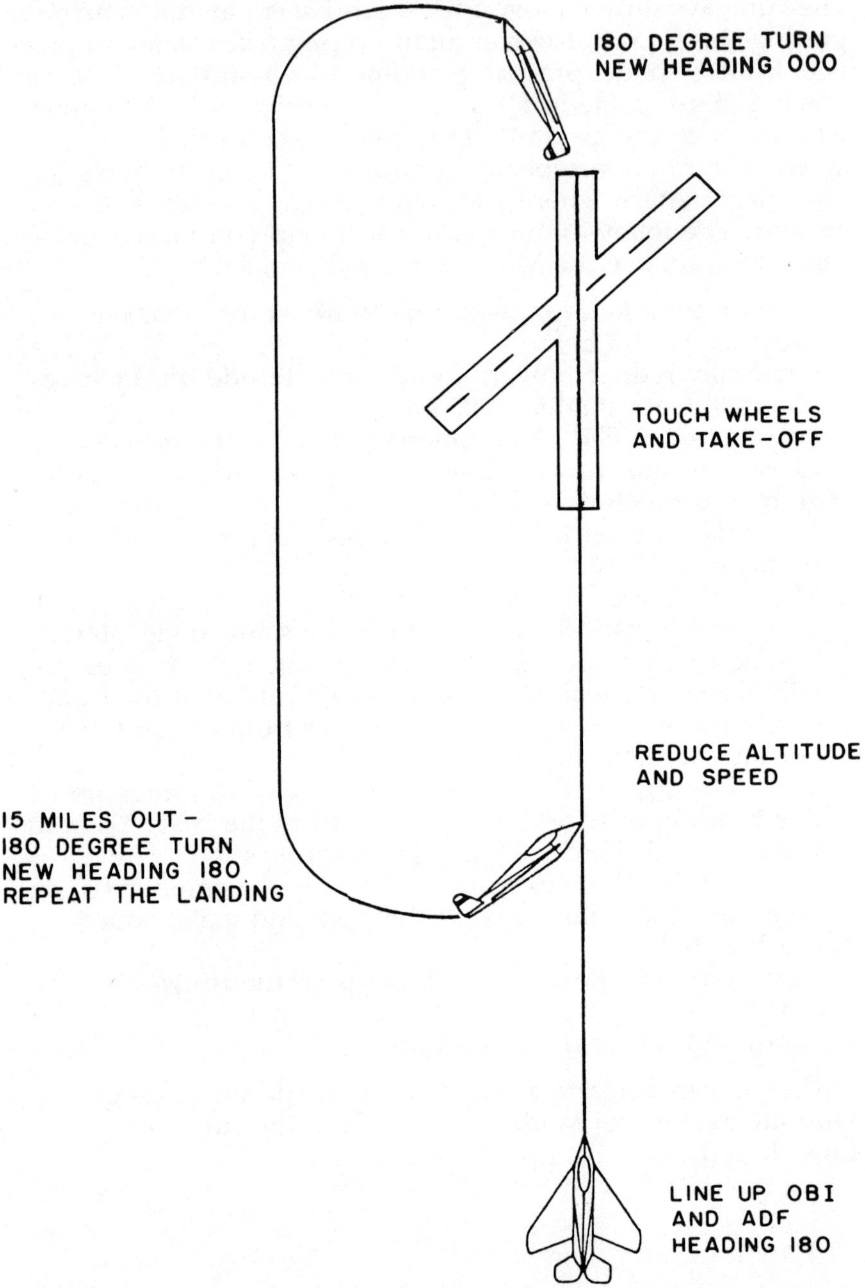

No Manual Tanking

Although it seems like a good idea, the engineers at SubLOGIC tell me it's impossible to hook up to the tanker manually. The SubLOGIC tanker is fast: The fuel transfer process takes place at mach 1.1!

Figure 8-8. Tanking at Mach 1.1

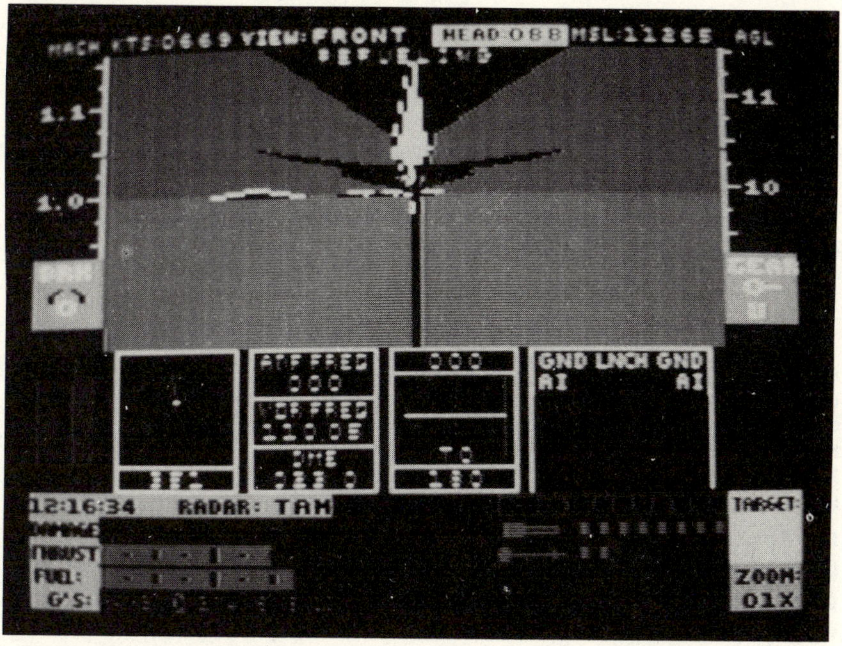

Scoring

Your final score is based solely on the number of targets destroyed in the current scenario. Skill level, hits taken, aircraft chosen, using automatic landing and tanking, and so on, have no effect on your score. There is no penalty for bailing out or crashing and there are no extra points given for successfully returning to the base.

Skill Levels

The skill level chosen will have an effect on three things: the maximum descent rate at which you can touch down without crashing; the number of hits you can take before being destroyed; and the distance at which you can be detected by enemy radar.

The only difference between regular, intermediate, and advanced missions is the number of targets and the type of targets. Weapons ranges, target vulnerability, and so on, are not affected.

Air-to-Air

The number of air-to-air combat encounters is purposely kept low in this game to allow you to concentrate on the strategy aspect of the game. They didn't intend to produce a fast action game. You may even begin to wonder why you bother to carry all of these air-to-air weapons.

In those rare cases where you run into enemy aircraft, you have a distinct advantage. Your AIM-7 Sparrows and AIM-9 Sidewinders have ranges of 40 and 10 miles respectively, while the enemy Atoll missiles can only reach out 5 miles.

A Bomb Is a Bomb

Views on this subject will vary from player to player. Some players insist on recreating the situations as realistically as possible while others try for high scores, using everything the game will allow. Those who value realism are offered a number of different air-to-ground weapons, each with its intended purposes. If realism is important to you, you should use Harpoons only on ships, cluster bombs only on runways, and so forth.

Those who are interested in using what the game provides to its maximum benefit should consider using Harpoon missiles on most ground targets. The Harpoon has the longest range and therefore the best standoff capability, which makes it the safest weapon to use. It is also extremely effective on all ground targets on land or sea. Maverick missiles would be a good second choice; they also have a good range.

CHAPTER 9
Tips and Hints for Playing *Stealth Fighter* by MicroProse

A new game from MicroProse is like a new Ludlum novel: always eagerly awaited and always received with enthusiasm. *Stealth Fighter* is no exception. Shortly after the project was announced, questions about the release date became common on the MicroProse phone lines. Players were not disappointed. The graphics and game play are excellent and *Gunship* pilots will recognize the popular promotion and decoration system. With some practice and a little help from the following hints you just might win that Congressional Medal of Honor.

The Big Hint

If you only remember one hint from this chapter, remember this one: Maintaining a low stealth profile is very important for a number of reasons. It improves your Stealth Rating and your score, and it reduces the chances of your being spotted by enemy radar. The best way to keep a low profile is to fly low and slow. Flight below 500 feet, where your profile is the lowest, however, can be very difficult. Constant altitude adjustments are necessary to keep from hitting the ground or popping up to 1000 feet.

Flight below 500 feet can be difficult, that is, unless you know the trick: There is a combination of airspeed and pitch that will allow you to fly hands-off at altitudes below 500 feet. This point will be different for each weapon's load and fuel level, but it is generally in the area of nine degrees of pitch and 250–275 knots.

The best way to find this combination is to fly at 900 feet and then set the pitch to nine degrees; this is indicated in the lower portion of the HUD (Heads-Up Display). Now decrease the throttle. Watch the Vertical Velocity Indicator bar gauge; when it indicates a slight descent, you're almost there. Now let the plane descend to below 500 feet. This should be a very slow descent; if not, increase your throttle by one notch. Once below 500 feet, you should have only one blue light showing on your EMV (Electro-Magnetic Visibility) bar. Increase throttle by one notch. This should stop your descent and leave you in level flight at 400 feet or so. Now if you want to change altitude you can increase or reduce throttle. If you can't fly steadily at this point and you reach a situation where one notch more of throttle causes the plane to rise while one notch less causes it to sink, adjust the pitch one notch up or down and repeat your throttle adjustment. This process becomes second nature with a little practice.

Once you master the technique of finding this smooth-flying slot, you'll be able to greatly increase your stealth percentage by flying consistently below 500 feet. Remember that you burn fuel faster at this altitude, so it may not be right for all missions. Also, be careful when flying in this manner

Figure 9-1. Flying "In The Groove" at 300 Feet

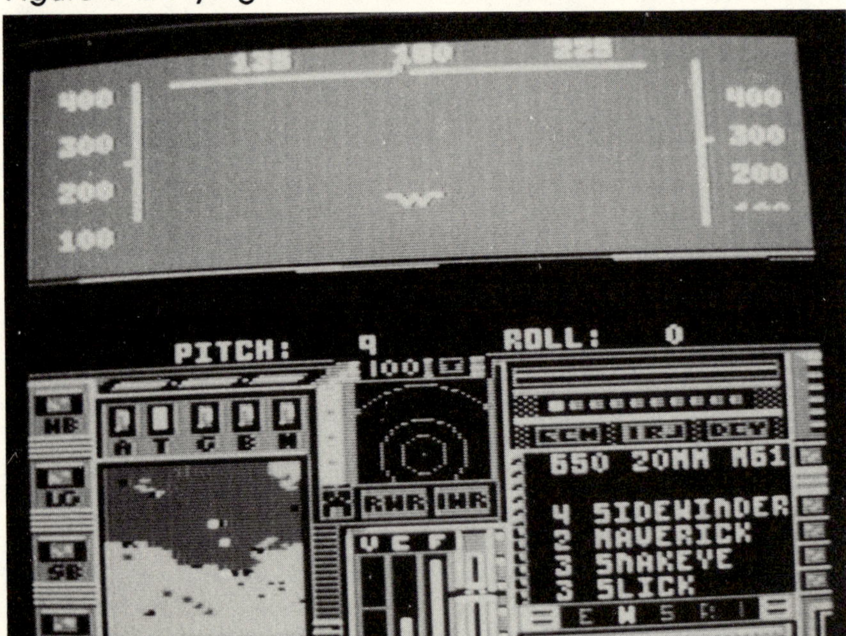

near mountains as the erratic winds near them will move you out of the slot and cause you to rise or sink. If you sink to 300 feet, there won't be a lot of air under you.

Radar Ranges

There is a drawing on page 42 of the manual that shows how to draw circles around all enemy positions to see the range of the radar at each installation. This would help you *thread the needle,* allowing you to wend your path just outside of radar range, thus minimizing your exposure. It isn't very clear how to do that, however.

Turn to page 64 in the manual. You'll see a SAM chart and under the heading *Max Range for Search* there is a number in parentheses. This number shows the radar range in numbers of blocks on the maps provided. If you knew that a SAM location had SA-2 missiles, you could go to the chart and find that the range was three. You could then use a compass to measure three blocks and draw a circle around the SAM site indicating the range of the radar. If you do this for all the radar installations in the area, you will be able to visualize the best route through them.

HARMS and Harpoons

Both HARM (High Speed Anti-Radiation Missile) and Harpoon missiles are excellent for accomplishing a specific mission: blasting radar and ships. The problem with these missiles is that you're limited to one missile in each weapons bay.

To earn a high score, you'll need to hit extra ground targets, but carrying one missile in each bay limits that ability. A good solution is to carry the multipurpose Maverick missile. The Maverick has a better range than the HARM and is just as effective. It does have a shorter range than the Harpoon, but you can safely get within Maverick range of most ships with little problem. The main benefit of the Maverick is that you can carry two missiles in each bay. This will allow you an extra missile for those important extra ground targets.

Weapons Are Internal

Remember that all the weapons carried by the F-19 are carried internally. When you activate these weapons and open the bay doors you increase drag and reduce lift. Get into the habit of increasing the throttle just before you activate your weapons. Doing so will prevent finding out the hard way that you are losing altitude.

This applies to dogfighting as well as to air-to-ground situations. If you're trying to out-turn a MiG, you're better off closing the bay doors until you're close to firing position. By doing so you will increase the turning performance of your aircraft and reduce the chances of stalling. You can use the air-to-air tracking mode to keep track of the enemy while keeping the doors shut; use the Switch Tracking Mode key to activate it.

Panic Shots

The enemy will often take shots with SAMs and AAMs (Air-to-Air Missiles) that don't have a chance of hitting you; they just try to trick you into using your ECM (Electronic CounterMeasures) systems and giving away your location. When you are entering or leaving an area with a good stealth profile, go to the 12-mile radar screen if shot at. Many times the missile won't even come within 12 miles of you, and if it does, its lock may be too weak to hit you. Wait until the last second before using ECM and use jammers only briefly. This will minimize your exposure.

Runway Landings

A common problem for beginning players is lining up the jet for runway landings. Most give up too soon. They come in low on fuel and have trouble getting properly lined up. When they give up and try to go around again, they run out of fuel.

You don't have to start your landing at the very end of the runway. The runways are very long and wide so you can be way out of line as you go over the start of the strip and still have time to straighten up and land on the last third of the runway. You'll be flying very slowly, so you have more time than you think. You'll be surprised at how fast the aircraft stops rolling after touchdown once you cut the engines and put on the brakes.

In fact, I don't recommend trying to land at the beginning of the strip; it's very hard to judge when you're over the runway and it's easy to land too short, thereby ruining an otherwise good mission. Give yourself plenty of room and set your plane down gently. You have plenty of time.

Enemy AWACS

Enemy AWACS (Airborne Warning And Control System) planes can cause a lot of problems in the North Cape and Central Europe scenarios. They can see you but you can't see them. If you've been spotted but you can't see anything on your radar, you've probably been seen by an AWACS. There isn't much you can do about AWACS except try to fly under their coverage.

Two-Player System

In the heat of battle, everything seems to happen at once. An extra pair of eyes and hands can be a big help. Many players have commented that they consistently score higher when they play with another experienced player at the keyboard. The following is one way to divide the responsibilities.

Pilot Responsibilities

Fly the aircraft
Set the course
Fire the weapons
Give the orders

Copilot Responsibilities

Change the radar display scale
Activate the ECM and decoys
Activate and select weapons
Keep an eye on the altitude during bomb runs and dogfights
Check all systems for damage after an enemy hit
ID and switch targets
Change map displays

Since the copilot is not involved with flying the plane, he or she can be given more duties to keep occupied.

Extra Fuel

Fuel is a critical item. You have to get to the target, destroy it, destroy extra ground or air targets, and return home. You'll rarely find this possible on the minimum amount listed as you arm your plane. If a mission has a minimum fuel amount of 9,900 pounds, you should think seriously about leaving some weapons behind and taking more fuel instead. Even then you will have to be frugal: The extra tanks don't carry very much. Be conservative with your throttle and altitude.

You can activate the fuel as soon as you take off. Doing so has the odd effect of increasing your speed and thus your range. The fuel is still there to use but you are no longer burdened by the extra weight. If you think you'll have to glide most of the way home from your mission, you might want to wait before using the extra fuel. Go ahead and let the aircraft run out of fuel (make sure that any extra weapons are dropped) and start your glide. Now activate the extra fuel, but don't start your engines yet. When you are close to the base, you can restart the engines and use the extra fuel for your landing maneuvers.

Gliding

If you don't turn or maneuver, you should be able to glide indefinitely under certain circumstances. It glides best when the weapons bays are empty. By pitching the nose up to around nine degrees you will be just above stall speed, and you should be able to get the Vertical Velocity Indicator to level out. This maneuver won't work if you have a lot of fuel left, due to the added weight.

Fuel Efficiency

The aircraft fuel consumption rate goes down as you increase in altitude, up to 32,000 feet. This is shown in an increase in airspeed for a particular throttle setting. The effect of this increase in airspeed is an increase in range.

Additional Ground Targets

Additional ground targets are essential for high scores. You'll receive the same score for any ground target.

If you have light weapons, SAM radar sites are excellent targets. Destroying this site with missiles or by strafing will disable all of the launchers. It's often a good idea to take out the SAM radar first, before attacking the major objective.

If you have heavy weapons, take out a runway. Enemy

planes will scramble from the closest runway. If the nearest runway is destroyed, they will have to come from the next closest base, which may be very distant. This should make your flight home a lot easier.

Enemy Fighters

Each type of enemy plane is programmed with different capabilities, as is each type of air-to-air missile. The airplane and missile attributes don't have as large an effect on the dogfighting capability as the quality of the enemy pilot. Green pilots will generally fly cautiously, keeping their planes flat and attempting few acrobatics. Average pilots will be somewhat more aggressive. Veteran pilots will use a variety of vertical maneuvers you won't be able to duplicate. Engaging a veteran enemy pilot in a close-in dogfight isn't advised. Even the folks at MicroProse have a hard time with them. You would be better off dealing with them at long range and then using stealth to escape.

The good news about enemy fighters is that they can use only missiles against you. They are not programmed for cannon attacks.

Also, it is to your advantage that enemy fighters are vectored to your last known position. They aren't vectored on an intercept course with your heading. They can be fooled if you let yourself be seen a distance from your target area. Once you are seen, the fighters will head for your current position. If you're not there when they arrive, and you havn't been sighted elsewhere, they will hang around for a few minutes and then go home. If you can get a low stealth profile, you can fly on toward your target without problems from enemy fighters.

Strafing Attacks

One way to increase your scores is to master the art of strafing targets. You carry all that cannon ammunition, so you might as well learn to use it. You must be below 500 feet for a successful attack, and your range is only 2.5 miles. Flying very slowly is the key. The manual recommends using the speed brake, but it's easier to use your flaps to increase lift and decrease stall speed. Line the target up and dip the nose slightly as you fire. Keep an eye on your altitude; making smoking holes in the ground won't win any medals.

Most targets will take several strafing hits to destroy. Some (runways, for instance) are almost impossible to destroy with your gun. Oil tanks are easily destroyed by gunfire.

Organize your Weapons Bay

If you take the time to organize your weapons before you take off, you can save precious seconds in the heat of battle. Some people like to put all air-to-air missiles in the top two bays and air-to-ground weapons in the bottom bays. Organize the way you like, but keep things the same so you'll know where they are at all times. You don't have time to pull up the weapons inventory in the middle of a dogfight.

Scoring System

As with all MicroProse games, the scoring system for *Stealth Fighter* is complicated. MicroProse tried to take into account all of the various options.

First you have a score based on the values of all the targets destroyed. That number is then multiplied by a factor. This factor value is based on

- ⊙ Region
- ⊙ Tension
- ⊙ Mission
- ⊙ Range
- ⊙ Enemy quality
- ⊙ Realism
- ⊙ Landing

Depending upon these items, the factor value may be larger or smaller than 1, thereby increasing or decreasing your basic score. If all options are at their hardest levels, your basic score can be increased by a factor of 9.

The following describes all items that go into making to the factor value. In order of importance they are

1. Region
2. Tension, enemy quality, and realism
3. Range
4. Mission type

Landing safely at a base will keep your score the same. Bailing out over the deep ocean costs you a few points, bailing out over friendly territory costs more, and bailing out over enemy territory (including coastal waters), will cost you heavily.

Let's examine each of the categories.

Region. Western Europe provides the most points. The North Cape provides only slightly less. These are followed by the Persian Gulf, Libya, and training.

Tension. In order of importance from highest tension to lowest, the levels of tension are Conventional War, Limited War, and Cold War.

Enemy Quality. The enemy may be a Veteran flier, a Regular, or Green (inexperienced).

Realism. The quality of realism will be either Realistic Flight, Easy Flight, or No Crash.

Range. The range is based on the Fuel Distance Estimate provided at the beginning of the mission. The range falls into one of four categories: 11,500 or more; 9,000 to 11,499; 7,000 to 8,999; and 6,999 or less.

Mission Type. There are also four levels of mission types: Ground Target; Aircraft Target; Air-to-Air Practice; Bombing Practice.

Promotions

Promotions are based on three things: number of missions flown; total score; and average score per mission. The average score prevents bad players from being promoted simply by flying a large number of missions.

To give some idea of the promotion process, MicroProse provided the information in Table 9-1. The numbers may not be exact. This table is only intended to provide an overall view.

Table 9-1. Factors Affecting Promotions

Rank	Number of Missions	Total Score	Average Score
First Lieutenant	2	400	150
Captain	5	1500	200
Major	10	3300	230
Lieutenant Colonel	20	7250	260
Full Colonel	40	16000	280
Brigadier General	99	22000	220

As you can see, it's important to keep your average score up. Don't record low-scoring missions. Also note that you must meet all of the criteria to be promoted. If you fly enough missions and achieve the required total score but your average is below par, you won't be promoted until you bring your average up.

Medals

Winning a medal is based solely on your performance on a particular mission; no other items are taken into account. Your score for that mission is the determining factor.

The values shown for earning medals may not be exact. They only approximate the points you must score.

The one-mission score values are shown in Table 9-2.

Table 9-2. One-Mission Score Values

Medal	Score
Congressional Medal of Honor	1800
Distinguished Flying Cross	1300
Silver Star	850
Bronze Star	500
Airman's Medal	250

To win a second medal of any type, you must score a little more than twice the original total. You'll win your first Silver Star before you win a second Bronze Star, and so on. To win your Third of any medal type requires more than three times the original total and so on. This keeps you from racking up 20 or 30 Airman's Medals without improving your flying.

IBM Version

By the time this book is published, MicroProse will have released the IBM version of *Stealth Fighter*. They have made use of the extra memory to improve the game so much that they felt a different title was warranted. Therefore, the title of the IBM version will be *F-19 Stealth Fighter*.

The cockpit is redesigned in several ways. The Stealth Profile indicator will be changed to let you know how close enemy radar is to locating you, and a Tracking Camera will be included. Once locked onto a target, this camera will provide a continuous view even if the target is behind you. This will be a big help in dogfights with other aircraft. The view from this camera will also show close-up shots of your missiles hitting the target, the smoking aircraft or ground target, and the enemy pilot's parachute.

CHAPTER 10
Tips and Hints for Playing *Jet* by SubLOGIC

Jet is the logical progression from SubLOGIC's popular Flight Simulator software, which simulated a Cessna aircraft. This exciting simulation of the F-16 Falcon and the F-18 Hornet is SubLOGIC's answer to the many requests for a game with more action. The flight characteristics of these aircraft were faithfully reproduced and, as such, *Jet* may require a little more time to master than other flight simulation programs. While it can be frustrating at first, with a little practice and some helpful hints, you'll be on your way to Ace status at mach speed.

Takeoff and Landing: Take it Easy

A common problem with first-time *Jet* pilots is they're too heavy with the controls. It's best to make changes a little at a time until you get the feel of the controls. You aircraft will not stop the maneuver as soon as you let go of the stick or neutralize your pitch, it will continue on for a second or two. You must anticipate this reaction and take the controls off early. Otherwise, you'll overcontrol, then overcompensate in the other direction as a reaction. Take it easy, think about what you want to do, and maintain a light touch on the stick.

Carrier Takeoffs

Trying to get the aircraft safely off of the carrier is one of the first problems players experience. Catapult launches can be tricky. The best way to master them is to establish a procedure and follow it every time you launch. The following procedure should get you safely into the air so you can get on to other things.

- ⊙ Apply full power and afterburner.
- ⊙ Press *L* to launch your aircraft.

⊙ When you reach a speed of 0.5 mach, apply two quick up-elevator keystrokes and neutralize your pitch. (If you are using a joystick, the pitch will automatically be neutralized when you let go of the stick.)
⊙ Raise your landing gear.

Carrier Landings

Carrier landings are difficult. You will have to practice the landings to master them. A few bits of information will make the process easier.

When trying to line up for a landing, use the lines on the water to help position yourself while you're still a good distance away from the ship. These lines run perpendicular to the landing strip on the ship and represent one-mile intervals. So, if you're flying across these lines and headed towards the ship, you are close to being correctly lined up. If you are flying down or with the lines, you are flying across the ship's path and you aren't lined up correctly for a landing. Also, you can use the magnified views to see how well

Figure 10-1. Flying Across the Lines Aligns Your Path with Carrier Flight Deck

you are lined up when far away from the ship (remember to go back to 1X to help judge your landing).

Don't make too big a deal about getting perfectly lined up either. You can hit the deck at any angle as long as you touch the white wire before you plunge off the deck or hit the tower. You don't even need to land from the right direction. You can come in on a heading of 270 and still land just fine. All you have to do is to land on the deck and then taxi over to the wire.

That brings up another point: If you miss the wire when landing, all is not lost. If you keep your speed brake out and cut your power you can still land on the deck, turn your plane around, and taxi back to the wire. So if you miss the wire, you don't have to go around for another landing attempt like a real pilot would. Simply put the plane down and taxi back to the wire.

Air-to-Air Combat: Offensive Maneuvering

The default weapons load for air-to-air missions is four Sparrows and four Sidewinders. Since you will probably fire several missiles before you start any heavy maneuvering, there is little harm in the extra weight of two more of each missile type. When the arming menu comes up, add two more Sparrows and two more Sidewinders.

The often-seen three-against-one scenario and the lack of countermeasures equipment (such as flares and chaff) limit the amount of ACM (air combat maneuvering) possible. Most of your time is spent trying to outmaneuver enemy missiles. Some tactics will improve your chances of making kills and surviving the fight. Theoretically, you should be able to get off the first shot. Your AIM-7 Sparrow missiles have more than twice the range of a AA-2 Atoll. But the game designers took away this advantage and programmed the Atoll with the longer range, so the MiGs will always shoot first.

You aren't limited to the range indicator. You may fire and score hits on targets before the indicator turns black. A good tactical plan when outnumbered is firing a salvo of four Sparrows before your indicator turns black. Then begin your evasive maneuvers to avoid the incoming missiles.

During your evasive maneuvering you should arm your AIM-9 Sidewinders for an opportunistic short range shot that may present itself.

Defensive Maneuvering

Defensive maneuvering is very important in this game since you aren't equipped with flares or chaff. Successful maneuvering is usually divided into two segments.

The first part of the maneuver involves flying at a right angle to the incoming missiles. This will usually cause them to fly behind you and then turn towards you. Once this happens, the second phase begins. At this point you should turn hard toward the missiles and start a spiraling climb that forces the missiles to constantly change direction and climb, thereby using their fuel. You must turn towards the missiles, not away from them. Turning away would only give them a better position behind you. You can mix a series of loops with the climbing turns, which may also cause the missiles to miss.

Use of your afterburners should be limited during defensive maneuvering because the added heat will only provide a stronger signal to the enemy heat-seeking Atoll missiles.

If you are hit, you can still attempt revenge. You will have several seconds of spinning before your plane breaks up. Take this time to arm and shoot all of your Sidewinder missiles if the enemy is close, or Sparrows if he is further away. When you "punch out," your missiles may still have time to find their mark and score another kill for you. Take care not to hang around in the cockpit too long. If you're still inside when the plane breaks up, you will lose your opportunity to use the rest of your allotted aircraft.

Air-to-Ground Attack Missions: Dive Bombing

The dive bombing technique seems to be the best technique in *Jet*. Though it does place you in a position to be fired at occasionally, you will score more hits than trying to pop up from a low level. The additional speed and energy picked up during your dive can be used to outmaneuver any missiles that may be fired at you.

When first learning to dive-bomb, you should follow the procedure outlined below. As your skill improves, you can attempt to drop bombs from lower altitudes, or from farther away, to reduce your exposure.

1. Launch, maintain your heading, and climb to 5000 feet. Make sure your view is set for 1 or 2×.

Figure 10-2. Evading an Incoming Missile
At position 1, you encounter an incoming missile. Turn to take a path at a right angle to the missile's path. This will generally cause the missile to pass behind you. At position 2, the missile is behind you. Break hard in the direction of the missile. Once again the missile should be forced to change directions and miss behind you. At position 3 start climbing turns. Turning away from the missile at position 2 (dotted lines) will only give the missile a better shot at you.

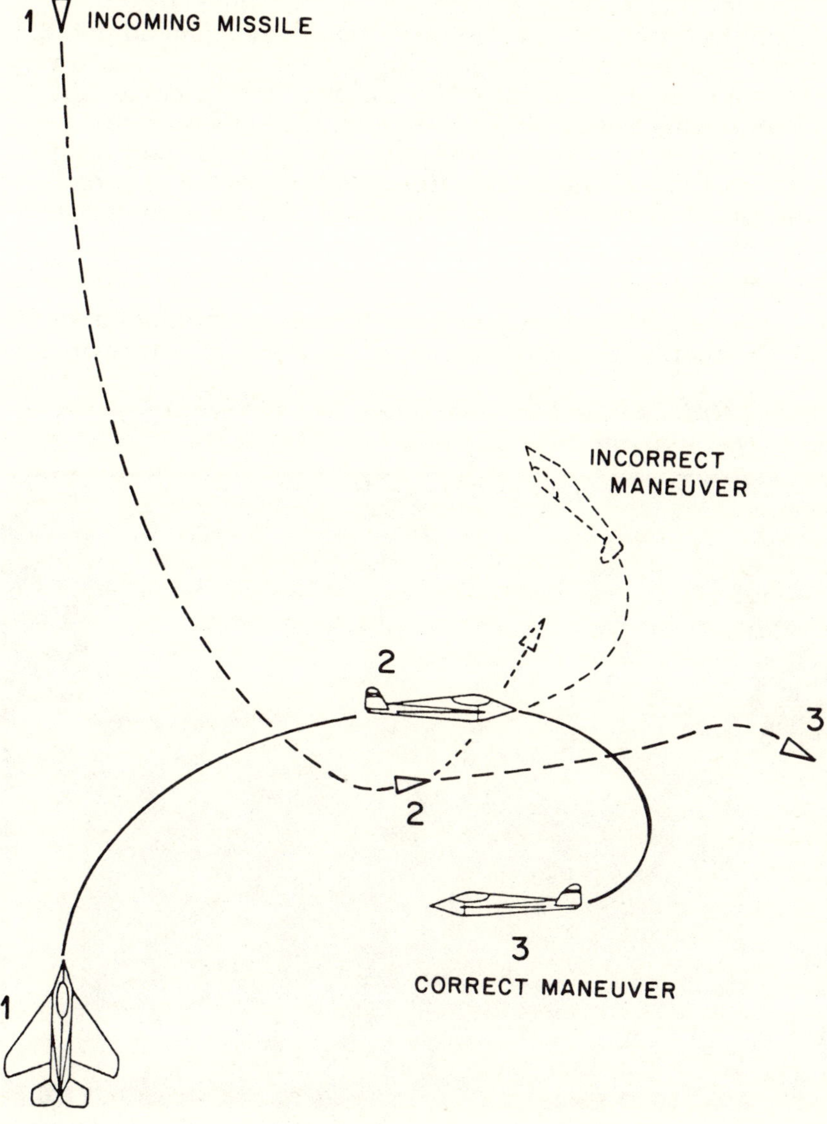

2. Once you reach 5000 feet, start a turn to your right; continue to climb.
3. A target should come into view shortly. Stop your climb at 10,000 feet and line the target up directly in front of you. Maintain level flight; cut your afterburners.
4. Watch your radar screen. When the target is halfway between you and the edge of the screen, start your dive. Center the target in the middle of the range circle.
5. Now set your view to 8× and make any small adjustments to center the target. When on 8 × view, make very small course corrections because any small change you make will cause a large change on the screen.
6. Drop your weapon as you pass through 5000 feet. Pull up, turn away from the target, and go back to 2× or 1× for maneuvering. If you're fired on, Start a hard spiral climb.

Magnified Views

It's very important that you learn to use the magnified view option when aiming air-to-ground weapons. If you're on 2×,

Figure 10-3. At 1× Magnification the Target Appears to Be in Your Sights

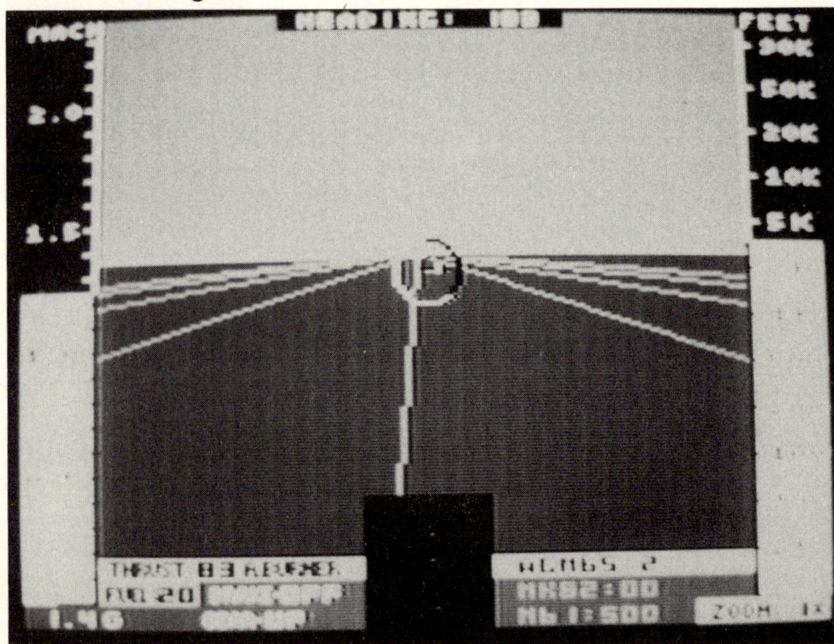

Figure 10-4. At 8 ×, You Can See Discrepancy

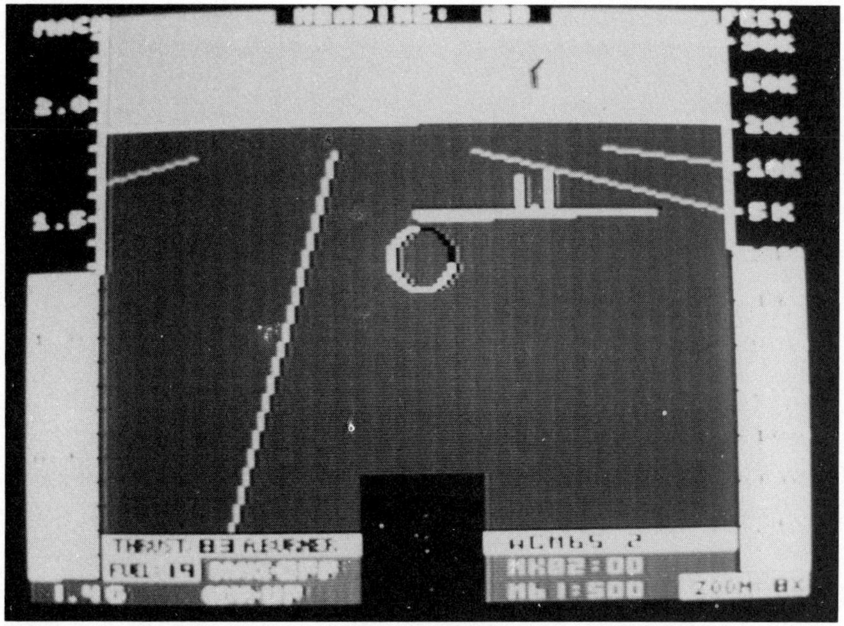

you may think the target is perfectly lined up, but by going to 8× you'll often find that the target isn't in the center of the circle and further minor adjustments are necessary.

Also, you should take care when maneuvering at high magnification levels. These levels will severely limit your field of vision and give you a distorted picture of what is happening; a shallow dive will look very steep at 8×.

Scenery Disks

Jet is compatible with the series of scenery disks offered by SubLOGIC. While you cannot use these disks for combat missions, they could be used to stage aerobatic routines at one of hundreds of airports across the country. With the San Francisco disk you could perform a show over the Golden Gate Bridge and Alcatraz Island.

These disks have one additional benefit: The fuel gauge will not work. You can fly for an unlimited period of time. While this might not be realistic, it can be a lot of fun as you cross the country, going from disk to disk in record time.

CHAPTER 11
Tips and Hints for Playing *Ace II* by UXB

Ace I was an arcade-style shoot'em-up that was fun to play but somewhat limited in its long-term appeal. With *Ace II,* UXB has found a way to keep the game fresh for an extended period. *Ace II* is one of the rare flight simulators that allows two players to fly against each other in a rather simple one-screen format. Each player is given half of the screen and a minimum of controls. The planes react quickly and the screen update speed is very fast.

Even with only a few controls, you can improve your scores and performance if you know certain facts about *Ace II*.

Aerobatics

Although you are limited to a front view only, this is a great game for learning and performing aerobatics. With both players looking at the same screen, communication is easy. Try some follow-the-leader maneuvers where one plane follows the other visually. Head-on passing maneuvers can also be performed, but remember to keep your distance. This game allows for midair collisions.

Chase Them Down

Computer pilots commonly try to lose the chasing plane by going into a hard dive. Unlike many other games, *Ace II* allows you to lose altitude very slowly. Don't be afraid to follow your opponent into a hard dive where you can't see the horizon. The programmers provided a pitch and roll display, so use it to keep track of where you are in relationship to the horizon. When you want to pull out of the dive, just level your wings and pull up.

Figure 11-1. Following the Leader, Inverted

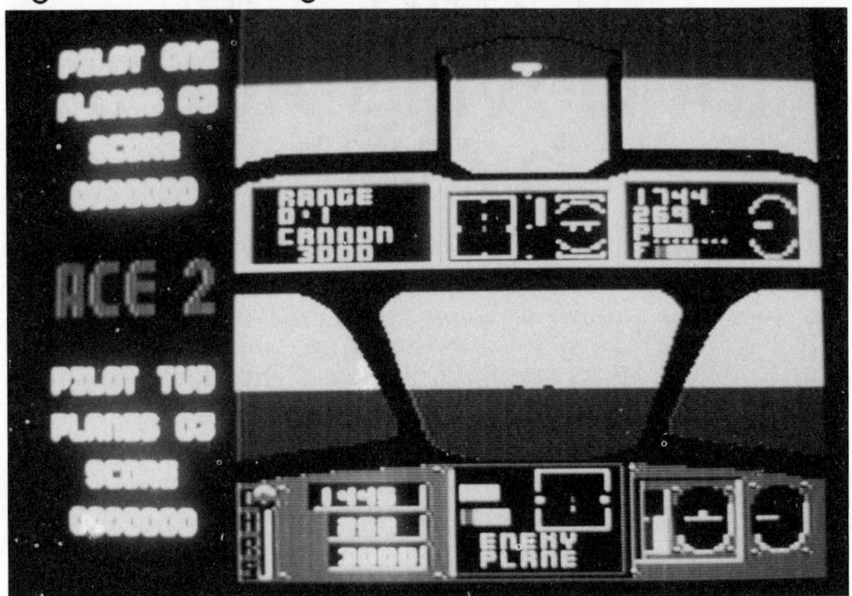

Know Your Opponent's Altitude

Changes in altitude can cause you to lose track of your opponent, so it's imperative that you keep an eye on both sets of gauges. The altitude of the enemy plane is displayed in the message area of your display but it only appears every few seconds. The best way to keep track of what's going on is to look at the other player's instruments.

Maximum Rate Turning

You must learn to use the throttle in your aircraft judiciously. If you go through the entire game at maximum speed with your afterburners lit, you will constantly overshoot your opponent and be out-turned. High speed maneuvers also cause you to run out of fuel prematurely. When you're involved in a tight dogfight, reduce your throttle so your speed is in the 150- to 250-knot range. This will allow your aircraft to turn at its maximum rate. Remember, if you want to climb hard at this throttle setting, you'll need to increase power or face the possibility of a stall.

Out of Ammunition

If you run out of ammunition, don't give up. Unless your opponent is very low also, you are probably going to lose your plane. But if you have to go, why not take him with you? Use the game's midair collision capability as a weapon: Try to ram your opponent. This can be an especially good tactic if you have superior numbers of aircraft.

Weapons Load

Ninety percent of this game is fought at close quarters. You will have few opportunities to use your long-range radar-guided missiles, so load up with eight heat-seeking missiles and only take four radar-guided missiles. This will give you more missiles than the enemy has flares; you should get some hits.

Avoiding Radar-Guided Missiles

The first stage of an engagement is usually an exchange of radar-guided missiles. Your first instinct is normally to turn hard and avoid the missile. This is not a good idea. For a radar-guided missile to be effective, the target must stay in the sight area for the entire flight or the missile will lose its lock. If you start maneuvering hard while still a long distance away from your opponent, he will be able to keep you in his sights very easily. He will fire radar missiles at you one after the other.

The best way to handle this situation is to get it over with as quickly as possible. Keep flying toward your opponent as fast as possible. You can use chaff and change altitude to avoid the missile. Once you get close, he will not be able to keep you in his sights and he'll lose lock-on. When this occurs, you have made the transition into a short-range dogfight engagement.

Fly High

Altitude is a very important component of air combat. Your computer opponent knows this and his first move will always be to fly straight up to 50,000 feet. If you don't follow him up to this altitude, he will have an advantage, so follow him. Level out as soon as you have achieved your altitude and start to look for your opponent. If you're still climbing with your nose pointing up when you come into radar range, he will fire the first shot because he can see you on his screen and you can't see him on yours.

Air-to-Ground Bombing and SAMs

The bombing portion of the game yields the most points and, as such, can be very difficult. The ground targets are heavily defended with surface-to-air missiles which are almost impossible to avoid. Flares and chaff seem to have little affect on them. Given this, it's best to ignore the missiles and go ahead with the bombing run, knowing you're going to get hit. The best way to do this is to:

- ⊙ Approach the target at top speed and below 1000 feet until you're fired on for the first time.
- ⊙ Once you are fired on, cut your speed to less than 500 knots. You can drop flares if you want, but they don't seem to have much effect.
- ⊙ When the target indicator box appears on the screen around the target, the target is in range. Immediately drop one of your bombs.
- ⊙ You must keep the target in the sight. If (or when) you are hit, regain control as quickly as possible to keep your target in the sight.
- ⊙ When your first bomb hits, drop another immediately.
- ⊙ As long as this second bomb hits the target before you take your third hit (I assume for purposes of this discussion that you use the *three missile to kill* setting), you will successfully complete the mission. The trick is to hit the target twice before it hits you three times.

Out of Fuel

As mentioned earlier, you will lose altitude slowly. This will allow you to glide for a long distance when out of fuel, so don't panic. You have to reach the edge of the map. When your fuel level becomes critical, gain as much altitude as possible with your last few drops and glide the rest of the way.

Note that if you are playing with the Crash Detection off, you can always make it back to the base. If you run out of fuel and cannot glide back to the base, just put it on the ground. You will continue on your present heading at whatever speed you were when you hit at altitude 0. It may take

Figure 11-2. Stalemate Turn

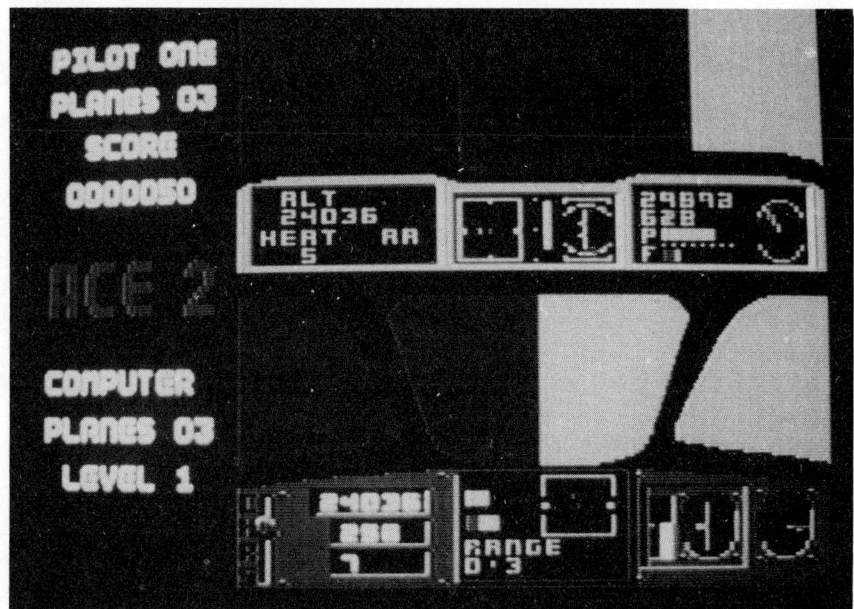

In a turning contest like this, no one will win. To break the stalemate you must maneuver in the vertical plane.

a while to get back to the base, but if you leave it alone long enough, you'll make it back. If you're playing against the computer in this mode it will take advantage of this, so you should learn to do it, too.

Maneuver Vertically Also

The two aircraft in this game are very similar in performance. If both players are at the same speed and bank angle trying to out-turn one another, neither will have an advantage and it will be a stalemate. One way to mix things up and get an occasional chance at a shot is to maneuver vertically as well as horizontally. Try the loop, dive, split-S, and Immelmann turn. Maneuvering in this manner will allow you to put some space between you and your opponent and give you time to line him up for a shot.

CHAPTER 12
Tips and Hints for Playing *Jet Combat Simulator* by Epyx

Jet Combat Simulator was one of the first games of this type (its copyright date is 1985), yet it has a number of good qualities that can provide a continuing challenge. The landing and takeoff sequences are very good. They are difficult to master, as they should be in an accurate simulation. The screen is rapidly updated, which provides for smooth flying and accurate maneuvers.

While the combat mode offers only guns (no missiles), the close-in knife-fight action can be very exciting. Beware of the higher skill levels: The enemy can pull some fancy tricks that might make you think the computer is cheating by not following the laws of physics.

Landing Practice

Takeoffs and landings are taken for granted in many games. In *Jet Combat Simulator,* they require practice to master. Before you go out hunting for MiGs, be sure you can get your bird back down on the ground safely. Take some time to go to the Practice Landing section of the game and learn the procedure. Once you can safely land and takeoff three times in a row you're ready to move on to combat.

When you master the following procedure for taking off and landing, you won't have to worry about getting killed while landing. This will allow you to concentrate on other aspects of the game.

Landings. To practice landing, select the first item on the game menu: Landing Practice. Once this is selected, the cockpit screen will appear and you will be properly lined up for a landing. To land safely, follow these steps:

1. Immediately add power until the indicator is a little more than halfway across the scale. Extend your flaps all the way.

2. Pitch down to start losing altitude.
3. Level off at 300 feet. As the landing ball on the LSI passes through the center of the scale, start a gentle decline at a pitch rate of approximately −10. If you descend below 50 feet before you reach the beginning of the runway, level off and then descend again when you reach the beginning of the strip.
4. When you touch down (there is no noise; watch for the altitude to go to zero) immediately cut the power to 0 and hold the brake key down until the plane comes to a stop. A menu should come up.
5. Select C for *continue.* The cockpit screen will now appear; you are on the runway where you stopped after your landing. You won't have enough runway in front of you to take off again. You'll have to taxi around to face the other direction.
6. To control your direction while taxiing on the ground, use the *Z* and *X* rudder-control keys. The runway isn't wide enough for you to turn around from your position in the middle. You'll first have to taxi to the edge. You will need some speed. Apply three keystrokes of throttle. Keep your speed down. When the speed level reaches 10, use the brake to slow it down occasionally. Hold down the *Z* key to point the aircraft toward the edge of the runway. Once you are close to the edge, hold down the *X* key until the plane is turned around. Line up the end of the runway on the center of the screen. The center isn't marked, but it's directly above the C in the word *PITCH* on the instrument panel. If you've made a good landing on the first part of the strip you'll want to taxi past the last set of dotted lines on the runway before making your turn; this will give you enough room to take off.

Figure 12-1. Runway Taxi Procedure
If you follow Path #1 (trying to turn around from the center of the runway), you'll run out of room and crash off the strip. However, if you follow Path #2 (taxiing to the side before turning around), you will have enough room to turn around for another takeoff.

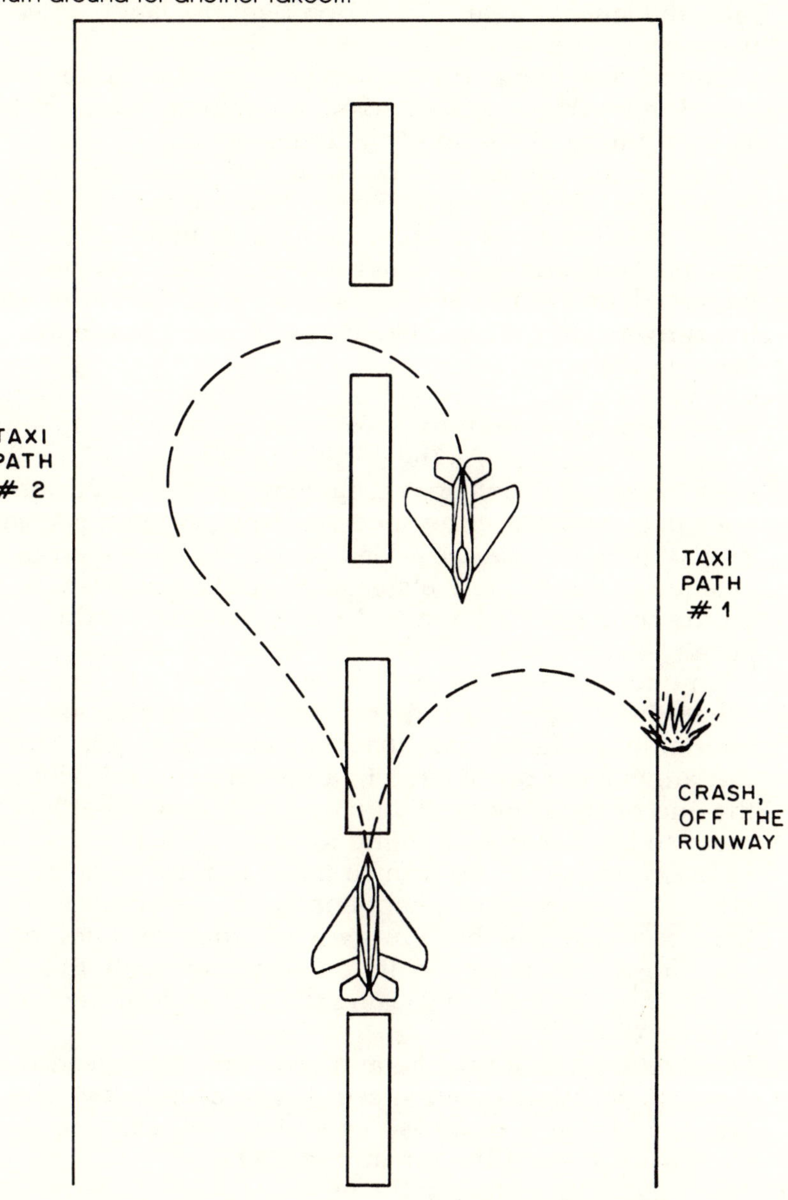

7. Once you have turned around, apply full power. When you reach 130 knots, pull back hard on the stick. The altitude indicator will tell you when you are airborne. Once up, immediately hit the *U* key several times to raise the landing gear. If you don't get the landing gear up quickly, you will crash.
8. Climb to 4000 feet and maintain your heading until you are 25 miles from the base. Note the mile indicator in the lower right-hand corner of the radar screen.
9. Once you are 25 miles away, make a 15-degree banked turn to the right until the diamond on the radar is even with the tail of the aircraft symbol. Now make a 23-degree banked turn in the other direction until your heading is 0 degrees. You should now be roughly in line with the runway; adjust your heading to point towards the flashing diamond.
10. Cut your throttle to just below the first major increment (approximately one-third power). Keep the nose up and reduce your speed. As the speed drops below 472 knots, start to extend your flaps. Each increment of extension of your flaps will cause the speed to drop. When the speed drops, more flap can be added. Once the flaps are fully extended, immediately increase power to past halfway on the scale to avoid a stall. Now lower your landing gear.
11. Pitch down to −40 on the VSI and level off at 500 feet.
12. When the ILS activates, fly towards the ball but maintain your altitude. Remember that the ball doesn't point to the runway but directs you to a line that extends from the end of the runway. You have to fly down that line in order to land properly. Flying towards the ball may, on occasion, cause you to turn so much that the runway is off of the screen. If this happens, be prepared to turn hard back towards the runway when the ball starts to move towards the center. If you don't anticipate this move you may end up chasing the ball back in the other direction.
13. Descend as you did before and level out at 100 feet, then down to 50. Watch your speed. If you drop below 130 knots, add a little power so you'll have enough speed to maneuver without the threat of a stall.
14. Touch down as before and stop.

Combat: The Bounce Attack

Since the earliest air combat of World War I, pilots have favored the *bounce attack*. The bounce attack is a maneuver in which you attack your opponent's tail from above. The name of this game could have been *Watch Your Six!* because most successful attacks are made from the rear.

The best way to attack your opponent is to take advantage of one of the game parameters. The result of this parameter is that if you and the enemy are separated by more than 5000 feet, he can't see you.

Here is the bounce attack, step by step:

- ⊙ Once your enemy is located, head straight for him and climb to 5000 feet above his altitude. The flight computer will give you the enemy altitude.
- ⊙ As you approach the enemy, reduce your speed to about half of the scale or lower.
- ⊙ When the bogey passes under you, turn and take a position behind and above him.
- ⊙ Make your bounce attack: Dive to gain speed and catch up with the bogey. When you begin to see the shape of the enemy aircraft, reduce your speed to avoid passing him. The worst thing you could do in this game is overshoot your opponent, thereby giving him the advantage. It's better to approach too slowly and let him pull away than to approach too fast and overshoot.
- ⊙ Wait until you are at close range to start firing. Your opponent will fly a zigzag course so don't over correct. If he is to one side, wait a second and he will likely reverse course and head back into your sights.
- ⊙ If you get too close to the enemy and you see the plane start to bank hard, you must react quickly. First, reduce your throttle to ⅓ of the scale or less; this will allow you to turn at your maximum rate. Second, turn hard with the bogey. If you can reduce your speed and turn hard enough, you'll be able to maintain your position behind the bogey; he may pull away a little, but you can always increase power to catch up.

Figure 12-2. When a Close Enemy Banks Hard, React Quickly to Avoid Overshooting

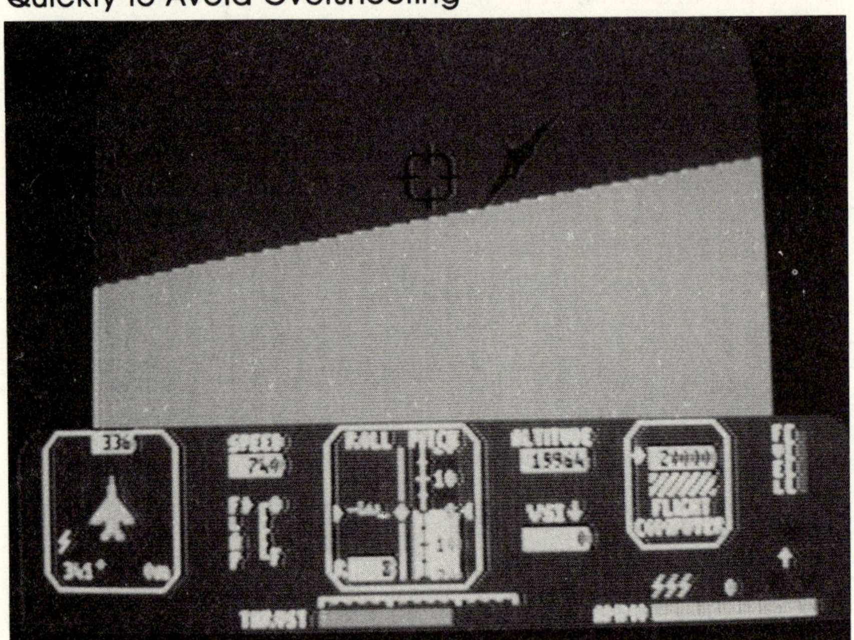

Shortcut

Often, after a long dogfight, you'll find yourself at the edge of the map. When the next bogey comes up, he may be all the way across the map. If you try to fly to him at top speed, you'll waste a lot of fuel and he will probably bomb a base before you get there. To get around this, fly off the map in the opposite direction from where you want to appear on the other side. For instance, if you fly off the top, you will reappear on the bottom. If you fly off of the right side, you'll pop up on the left side. Using this shortcut can save enormous amounts of time and fuel.

CHAPTER 13
Tips and Hints for Playing *Chuck Yeager's Advanced Flight Trainer*

Chuck Yeager's Advanced Flight Trainer, distributed by Electronic Arts, is one of the few jet fighter games in which you aren't expected to kill an enemy or bomb a target. There are plenty of things to do in this simulation, and plenty of planes to do them with. That's the appeal of this game: You can choose to fly a Cessna 172, the high speed SR-71, or one of 12 others in between, including birds from both World Wars.

With no combat options, the programmers had plenty of memory to concentrate on the simulation aspect and the flight characteristics of the various planes available. This also allows you to concentrate on your flying and aerobatic skills. *Advanced Flight Trainer,* or *AFT* as it is known, is not as simple to fly as some of the combat simulators and, as such, will require a bit more practice to master. What follows are a few tips to help get you started.

The Firebutton

Using the cursor on the Heads Up Display can be a bit confusing at first and you can quickly find yourself in some unusual positions. If things start to get out of control, you can press your joystick firebutton or the space bar to neutralize all of the control surfaces. Once you've done this you can use the Attitude Indicator to level your wings and the nose.

Flight Instruction

If you're new to this game, the best way to become acquainted with all the controls and the reaction of the aircraft is to go through the Flight Instruction portion of the soft-

Figure 13-1. It Would Be Nice to Get a Kind Word from Chuck Every Now and Then

ware. This section teaches you the interrelation between attitude and throttle setting that can be important in this simulation.

Center the Stick

Unlike most of the combat games, AFT requires that you center the joystick after a maneuver. With most of these games, the joystick will center itself after you move it. For example, if you release the stick in a 30-degree banking turn, the plane will stay in the turn. With this game, if you get into a turn and let go of the stick, the control surfaces on the aircraft will remain in the turning position. The aircraft will continue to roll until you apply an equal amount of pressure in the opposite direction to neutralize the surfaces.

High-G Maneuvers

Another realistic feature of this simulation is its reaction to high G loads. You need to keep an eye on the G-force indicator because a blackout or red-out will occur if you exceed certain limits. Remember that the aircraft can stand more stress than the human body.

Various Planes

Each plane will react differently. Take time to get the *feel* of a plane in the Test Flight mode before you get into racing or aerobatics. Even if you have flown a particular plane before, you may still need a quick orientation flight to get the feel of it again.

Racing

Each race course will require different skills to master. It's best to start with the straight course. This course is roughly straight, though you'll need to make some small maneuvers. It's best to line yourself up with the first gate while taxiing for takeoff. The P-51 will tend to pull to the right. You can compensate with a couple of hits on the rudder.

Once you learn to take off and fly through the gates in the straight course, you can take a shot at the two-mile box.

Figure 13-2. If You Think You're Good, Try Taking One of These Gates Inverted

Though they call this course a *box,* all of the gates are lined up north and south. Once you're through the first two gates, you'll have to swing wide to enter the next pair heading due south. This will require a 180-degree turn. You can use the left view feature to keep an eye on the course as you turn.

When passing through a gate, don't react too quickly to the message that says you are through the gate. This seems to come up a little early. You can still crash into the gate for a second or two after you see this message, so avoid any quick maneuvers.

The obstacle courses require excellent turning skills. You'll have to master hard turns at low altitude. The ground comes up quickly. Although you want to win the race, speed is not always the answer. Flying at 60 or 70 percent of your maximum speed will let you maneuver more closely to the obstacles, which will keep you from having to make such long hard turns. At higher speeds your turn radius will increase, forcing you to fly farther to turn.

Record Your Aerobatics

One of the best features of this simulation is the ability to record your own aerobatic routines to watch again and try to follow. If you developed a routine in the first part of this book, here's the place to try it out. The record feature works well once you realize they left one small bit of information out of the Commodore 64/128 instruction book.

Look at your disk. It's write-protected. There is no notch. To record your aerobatic flights and races, you must notch the disk. Standard notches are available that will do the trick. Be careful not to bend the disk or touch the recording medium through the slot.

To view a recorded flight, go to the menu, select the correct slot, and press any key but *Y* when asked if you want to record. The flight will then start. You will see your plane flying your previous routine, trailing smoke balls. Try to follow to get a score or just fall back a bit and watch. You can always zoom in if the plane starts to get out of sight.

Remote Control

The tower view allows you to fly your plane as if by remote control. You can either start on the ground or use the menu to position the aircraft lined up for landing.

You will be better off using one of the slower planes until your skills are sharpened in this mode. The jets tend to react very quickly and without the instruments to guide you, you're often one step behind the aircraft. The slower prop-driven aircraft give you more time to adjust your controls. They are also easier to keep in the area of the tower, so highly magnified views are not necessary.

Figure 13-3. Remote Control Flying from the Tower

Altitude Records

The maximum altitude limits listed in the manual are by no means the limits of what can be achieved. My manual lists the maximum altitude record for the SR-71 to be 164,900, but I've heard of others reaching more than 190,000 feet. I have had the F-18 well above 90,000 feet. How high can you fly?

Figure 13-4 illustrates how you can achieve greater altitudes.

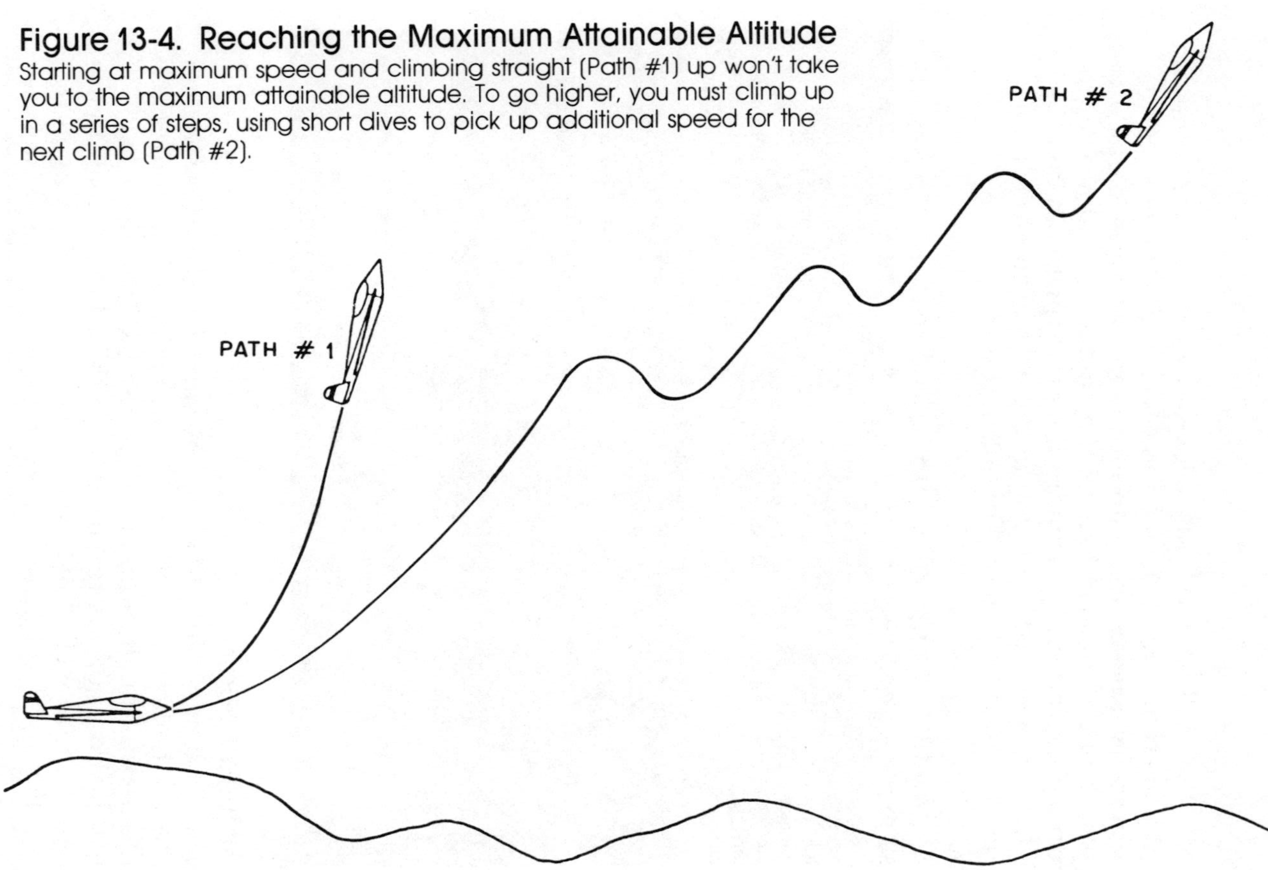

Figure 13-4. Reaching the Maximum Attainable Altitude
Starting at maximum speed and climbing straight (Path #1) up won't take you to the maximum attainable altitude. To go higher, you must climb up in a series of steps, using short dives to pick up additional speed for the next climb (Path #2).

Index